BI 3143840 7

KU-504-837

BIRMINGHAM CITY
UNIVERSITY
DISCARDED

Environmental Risk Harmonization

Ecological and Environmental Toxicology Series

Series Editors

Jason M. Weeks
Institute of
Terrestrial Ecology
Monks Wood, UK

Sheila O'Hare
Scientific Editor
Hertfordshire, UK

Judith Zelikoff
Nelson Institute of
Environmental Medicine
Tuxedo, NY, USA

The fields of environmental toxicology, ecological toxicology and ecotoxicology are rapidly expanding areas of research within the international scientific community. This explosion of interest demands comprehensive and up-to-date information that is easily accessible both to professionals and to an increasing number of students with an interest in these subject areas.

Books in the series will cover a diverse range of relevant topics ranging from taxonomically-based handbooks of ecotoxicology to current aspects of international regulatory affairs. Publications will serve the needs of undergraduate and postgraduate students, academics and professionals with an interest in these developing subject areas.

The Series Editors will be pleased to consider suggestions and proposals from prospective authors or editors in respect of books for future inclusion in the series.

Forthcoming titles in the series

Ecotoxicology of Wild Mammals
Edited by Richard Shore and Barnett Rattner (ISBN 0 471 97429 3)

Handbook of Soil Invertebrate Toxicity Tests
Edited by Hans Løkke and C. A. M. van Gestel (ISBN 0 471 97103 0)

Pollution Risk Assessment and Management: A Structured Approach
Edited by Peter E. T. Douben (ISBN 0 471 97297 5)

Statistics in Ecotoxicology
Edited by Tim Sparks (ISBN 0 471 96851 X cloth, ISBN 0 471 97299 1 paper)

Environmental Risk Harmonization

Federal and State Approaches to Environmental Hazards in the USA

Edited by

Michael A. Kamrin
Institute for Environmental Toxicology
Michigan State University
East Lansing, Michigan, USA

JOHN WILEY & SONS

Chichester · New York · Weinheim · Brisbane · Singapore · Toronto

Copyright © 1997 John Wiley & Sons Ltd, except those contributions which are identified as US or
        Canadian Government works
        John Wiley & Sons Ltd,
        Baffins Lane, Chichester,
        West Sussex PO19 1UD, England

        National   01243 779777
        International (+44) 1243 779777
        e-mail (for orders and customer service enquiries): cs-books@wiley.co.uk
        Visit our Home Page on http://www.wiley.co.uk
                    or http://www.wiley.com

All rights reserved. No part of this publication may be reproduced, stored in a retrieval system, or
transmitted, in any form or by any means, electronic, mechanical, photocopying, recording,
scanning or otherwise, except under the terms of the Copyright Designs and Patents Act 1988 or
under the terms of a licence issued by the Copyright Licensing Agency, 90 Tottenham Court Road,
London W1P 9HE, UK, without the permission in writing of the publisher.

This book is designed to provide accurate and authoritative information in regard to the subject
matter covered. It is sold with the understanding that statements and opinions expresssed are those
of the editor or individual contributors, and do not represent official statements or expressions of
policy by the organizations to which individual contributors are affiliated.

Other Wiley Editorial Offices

John Wiley & Sons, Inc., 605 Third Avenue,
New York, NY 10158-0012, USA

WILEY-VCH Verlag GmbH, Pappelallee 3,
D-69469 Weinheim, Germany

Jacaranda Wiley Ltd, 33 Park Road, Milton,
Queensland 4064, Australia

John Wiley & Sons (Asia) Pte Ltd, Clementi Loop #02-01,
Jin Xing Distripark, Singapore 129809

John Wiley & Sons (Canada) Ltd, 22 Worcester Road,
Rexdale, Ontario M9W 1L1, Canada

UNIVERSITY OF
CENTRAL ENGLAND

Book no.   31438407

Subject no. 363.70560973| KaM

INFORMATION SERVICES

Library of Congress Cataloging-in-Publication Data

Environmental risk harmonization : federal and state approaches to
    environmental hazards in the USA / edited by Michael A. Kamrin.
        p.   cm. – (Ecological & environmental toxicology series)
    Includes bibliographical references and index.
    ISBN 0-471-97265-7 (alk. paper)
    1. Environmental risk assessment–United States. 2. Environmental
management–United States. I. Kamrin, Michael A. II. Series.
GE150.E58   1997
363.7'056'0973–dc21   97–16265
                                                                  CIP

British Library Cataloguing in Publication Data

A catalogue record for this book is available from the British Library

ISBN 0 471 97265 7

Typeset in 10/12pt Garamond from authors' disks by Acorn Bookwork
Printed and bound in Great Britain by Bookcraft (Bath) Ltd, Midsomer Norton, Somerset
This book is printed on acid-free paper responsibly manufactured from sustainable forestation,
for which at least two trees are planted for each one used for paper production.

Contents

Publisher's note The Editor and the Publishers wish to make it clear that the Appendices within this book are comprised of excerpts from the actual titled documents and as such are not intended to be complete.

List of contributors

Nicholas D. Anastas
Office of Research and Standards, Massachusetts Department of Environmental Protection, One Winter Street, Boston, MA 02108, USA

Richard A. Becker
Director, Office of Environmental Health Hazard Assessment, California Environmental Protection Agency, 301 Capitol Mall, Second Floor, Sacramento, CA 95814, USA

David A. Bennett
Senior Process Manager for Risk, Office of Emergency and Remedial Response (5201G), US Environmental Protection Agency, Washington, DC 20460, USA

P. Michael Bolger
Supervisory Toxicologist, Contaminants Branch (HFS-308), Center for Food Safety and Applied Nutrition, US Food and Drug Administration, 200 C Street, SW, Washington, DC 20204, USA

Christine F. Chaisson
President, TAS, Inc., 4350 Fairfax Drive, Arlington, VA 22203, USA

William E. Cooper
Institute for Environmental Toxicology, Michigan State University, C231 Holden Hall, East Lansing, MI 48824-1206, USA

Jeffrey A. Crum
Senior Toxicologist, Environmental Response Division, Michigan Department of Environmental Quality, PO Box 30426, Lansing, MI 48909, USA

Charles P. Cubbage
Pesticide and Plant Pest Management, Michigan Department of Agriculture, PO Box 30017, Lansing, MI 48909, USA

Christopher T. De Rosa
Director, Division of Toxicology, Agency for Toxic Substances and Disease Registry, 1600 Clifton Road NE, Mailstop E-29, Atlanta, GA 30333, USA

Michael L. Dourson
Director, Toxicology Excellence for Risk Assessment, 4303 Kirby Avenue, Cincinnati, OH 45223, USA

William H. Farland
Director, National Center for Environmental Assessment, US Environmental Protection Agency, 401 M Street, SW, Washington, DC 20460, USA

Maria Gomez-Taylor
Office of Water, Health and Ecological Criteria Division, US Environmental Protection Agency, 401 M Street, SW (4304), Washington, DC 20460, USA

Robert L. Griffin
Acting Chief, Planning and Risk Analysis Systems, Animal and Plant Health Inspection Service, US Department of Agriculture, 4700 River Road, Unit 117, Riverdale, MD 20737-1228, USA

John L. Hesse
Division of Health Risk Assessment, Michigan Department of Community Health, 3423 N. Martin Luther King, Jr. Boulevard, Lansing, MI 48909, USA

Kathy Hughes
Priority Substances Section, Environmental Substances Division, Environmental Health Centre, Tunney's Pasture, Ottawa, Ontario, Canada K1A 0L2

Harold Humphrey
Division of Health Risk Assessment, Michigan Department of Community Health, 3423 N. Martin Luther King, Jr. Boulevard, PO Box 30195, Lansing, MI 48909, USA

Barry L. Johnson
Agency for Toxic Substances and Disease Registry, Assistant Administrator, Assistant Surgeon General, 1600 Clifton Road NE, Mailstop E-29, Atlanta, GA 30333, USA

Michael A. Kamrin
Institute for Environmental Toxicology, C-231 Holden Hall, Michigan State University, East Lansing, MI 48824, USA

Thomas A. McDonald
California Environmental Protection Agency, Office of Environmental Health Hazard Assessment, 301 Capitol Mall, Second Floor, Sacramento, CA 95814, USA

Bette Meek
Head, Priority Substances Section, Environmental Substances Division
Environmental Health Centre, Tunney's Pasture, Ottawa, Ontario, Canada K1A
0L2

Jennifer Orme-Zavaleta
National Health and Environmental Effects Research Laboratory, US
Environmental Protection Agency, Mail Code MD 51A, Research Triangle Park,
NC 27711, USA

James N. Seiber
University Center for Environmental Sciences and Engineering, University of
Nevada, Reno, NV 89507, USA

Kirpal S. Sidhu
Division of Health Risk Assessment, Michigan Department of Community
Health, 3423 N. Martin Luther King, Jr. Boulevard, PO Box 30195, Lansing, MI
48909, USA

Cynthia Sonich-Mullin
International Programme on Chemical Safety, World Health Organization, 20
Avenue Appia, CH-1211, Geneva 27, Switzerland

Robert C. Spear
Department of Environmental Health Sciences, School of Public Health,
University of California, Berkeley, CA 94720, USA

Yee-Wan Stevens
Environmental Health Scientist, Division of Toxicology, Agency for Toxic
Substances and Disease Registry, 1600 Clifton Road NE, Mailstop E-29, Atlanta,
GA 30333, USA

James W. Stratton
Deputy Director, Department of Preventive Services, California Department of
Health Services, P.O. Box 942732, Sacramento, CA 94234, USA

Carol Y. Swinehart
Michigan Sea Grant, Michigan State University, 323 Natural Resources Building,
East Lansing, MI 48824, USA

David T.-W. Ting
Office of Environmental Health Hazard Assessment, California Environmental
Protection Agency, 301 Capitol Mall, Second Floor, Sacramento, CA 95814, USA

Lauren Zeise
Office of Environmental Health Hazard Assessment, California Environmental
Protection Agency, 2151 Berkeley Way, Annex II, Berkeley, CA 94704, USA

Foreword

It is difficult to keep pace with the continuously developing field of ecotoxicology; no general text or single book is able to meet the requirements of such a diverse subject. Consequently, two of the primary objectives of our new book series on Ecological & Environmental Toxicology published by John Wiley & Sons will be to address selected new topics and to identify and address insufficiencies in the current literature.

This incisive new book, *Environmental Risk Harmonization* edited by Michael A. Kamrin, the first to be published in the series, will provide an invaluable source of information for all those involved in environmental risk assessment, irrespective of their scientific background. This book addresses for the first time the internal conflict between risk assessment and risk management among and between government units, and how the use of science in public policy has created the uniquely interesting and developing field of risk harmonization. This is the result of the growing dissatisfaction and concern amongst scientists at the misinterpretation of the raw data used in the managerial and legislative processes that ultimately lead to policy decisions which in turn affect society.

The editor and contributors to *Environmental Risk Harmonization* present a clear and authoritative account of this developing field, and enable provocative insight into the relevance of the work in a broader perspective.

We hope that this book, the first of many to be published in the series, will prove a valuable and contemporary source of information for all those involved in the science of ecological and environmental toxicology.

Jason M. Weeks
Sheila O'Hare

Preface

This book is the outgrowth of the long-term interest of the editor in the use of science in risk assessment and risk management. It started 30 years ago with the search for examples that could be used to help undergraduates with little or no science background understand what science is and how it is used in making decisions in our society. The example that seemed most appropriate at that time was the management of human health risks from artificial sweeteners. About this time, cyclamates were banned as food additives in the United States and the scientific data used as the bases for the ban were published in a *Science* magazine article. Assessing the validity of the data and conclusions provided an opportunity to examine how scientific causation is established and what part science and scientists play in societal decisions about potential risks to human health from chemicals in the environment.

The artificial sweetener situation was a very fortuitous example as it grew in scope as further studies suggested that there were also risks from saccharin, the other available artificial sweetener. It also grew in complexity as different countries evaluated the same data and made different decisions about the seriousness of the risks from the various sweeteners. For example, Finland first followed the United States lead and banned cyclamates but changed this decision after a few years and allowed them back on the market. The United States and Canada adopted opposite approaches; cyclamates are banned in the United States and allowed in Canada while saccharin is banned in Canada and allowed in the United States.

As risk issues gained in importance during the 1970s and 1980s and many different laws and regulations were adopted, situations where different risk management decisions, based on the same data, were made by different government agencies became more common. A very good illustration of this occurred in the Great Lakes where the published estimates of risks from eating fish often depended on which state's waters the angler was fishing in rather than the level of contaminants in the fish.

In the 1990s, the implications of these disparities in the use of science became more and more apparent as a result of increasing international trade and the formation of new political entities, such as the European Community. Further, scientists became more aware of the way that their data were being used and felt the need to have greater input into the risk assessment and risk management processes, often because they felt that their data and conclusions were being misinterpreted.

Thus, the editor's interest in and research on the use of science in public policy concerning environmental chemicals evolved into an interest in risk harmonization; particularly, the bases for disparities in risk assessment and risk management among and between government units. To shed more light on this topic, the editor applied for and received funding to hold a conference and workshop on the *Harmonization of State/Federal Approaches to Environmental Risk*. This meeting was held on May 20–21, 1996 at the Kellogg Center at Michigan State University and brought together representatives of state, national, and international agencies responsible for environmental risk assessment and management as well as other interested parties.

The presentations made at the conference and the discussions held during meeting workshops form the substance of this book. The editor has tried to provide an historical overview in the first chapter and a summary of the consensus reached at the meeting in the fifteenth and last chapter.

While the many difficulties in achieving harmonization were clear during the proceedings, the meeting participants provided much encouragement that risk harmonization can and will be improved in the years ahead. This encouragement was accompanied by a number of specific suggestions as to directions to take and how best to move in these directions.

The smooth planning and execution of the meeting and the timely development of the manuscript for this book would not have been possible without the tireless efforts of Jon MacDonagh-Dumler. He has expertly played many of the different roles needed to complete a project such as this successfully: project manager, meeting facilitator, publicist, supervisor of student help, subeditor, graphics developer, and researcher.

The editor would also like to thank all of the presenters, workshop facilitators, planning committee members, and meeting participants for contributing their time and effort to this project. In addition, the staff of the Michigan State University Institute for Environmental Toxicology provided critical support services during this project. In particular, Carole Abel answered numerous phone calls and word processed the many versions of each manuscript. Carol Chvjoka managed to keep the books balanced and the auditors at bay and successfully accommodated the often unique funding requests that arise as part of meetings such as this. Darla Conley was also very instrumental in the successful completion of many facets of this project.

Funding for this project was made available through a grant from the Office of the Great Lakes, Michigan Great Lakes Protection Fund (MGLPF). In addition, the editor would like to thank the Michigan State University Provost and the Vice President for Research and Graduate Studies for their financial support to supplement the grant monies provided by the MGLPF. The in-kind support provided by the Director of the Institute for Environmental Toxicology, Dr. L. J. Fischer, is also much appreciated.

Michael A. Kamrin
East Lansing, Michigan, October 1997

Acronyms

ORGANIZATIONS

APHIS	Animal and Plant Health Inspection Service
ASTM	American Society for Testing and Materials
ATSDR	Agency for Toxic Substances and Disease Registry
Cal/EPA	California Environmental Protection Agency
CENR	Committee on Environment and Natural Resources
CEQ	Council on Environmental Quality
DEQ	Michigan Department of Environmental Quality
EPA	US Environmental Protection Agency
ERD	DEQ Environmental Response Division
FAO	Food and Agriculture Organization
FCCSET	Federal Coordinating Council for Science, Engineering and Technology
FDA	US Food and Drug Administration
FSTRAC	Federal-State Toxicology and Regulatory Alliance Committee
HARP	Health Activities Recommendation Panel
IARC	International Agency for Research on Cancer
IFCS	International Forum on Chemical Safety
IJC	International Joint Commission
ILO	International Labor Organization
IOMC	Inter-Organizational Programme for the Sound Management of Chemicals
IPCS	International Programme on Chemical Safety
IRLG	Interagency Regulatory Liaison Group
MDA	Michigan Department of Agriculture
MDCH	Michigan Department of Community Health
MDNR	Michigan Department of Natural Resources
MHPF	Minority Health Professions Foundation
NAS	National Academy of Sciences
NRC	National Research Council
NSTC	National Science and Technology Council
NTP	National Toxicology Program
OECD	Organization for Economic Cooperation and Development
OMB	Office of Management and Budget
OSTP	Office of Science and Technology Policy

OTA	Office of Technology Assessment
SSARP	Substance-Specific Applied Research Program
TASARC	Tri-Agency Superfund Applied Research Committee
UNCED	United Nations Conference on the Environment and Development
UNCETDG	United Nations Committee of Experts on Transport of Dangerous Goods
UNEP	United Nations Environment Programme
UNIDO	United Nations Industrial Development Organization
USDA	US Department of Agriculture
WHO	World Health Organization

RISK ASSESSMENT AND RISK MANAGEMENT TERMS

AOC	area of concern
ADI	acceptable daily intake
ARAR	applicable or relevant and appropriate requirement
CICAD	Concise International Chemical Assessment Document
CREG	cancer risk evaluation guide
DNA	deoxyribonucleic acid
EMEG	environmental media evaluation guide
EPI	Exposure Potency Index
HACCP	Hazard Analysis of Critical Control Points
HEAST	Health Effects Assessment Summary Tables
HQ	Hazard Quotient
HSDB	Hazardous Substances Data Bank
HPV	health protection value
IF	interaction factor
IPM	Integrated Pest Management
IRIS	Integrated Risk Information System
LOAEL	lowest-observed-adverse-effect-level
MCL	maximum contaminant level
MRL	minimal risk level
NOAEL	no-observed-adverse-effect-level
PAH	polycyclic aromatic hydrocarbon
PBPK	physiologically-based pharmacokinetic modeling
PCB	polychlorinated biphenyl
PRG	Preliminary Remediation Goal
PSL	Priority Substances List
RA	risk assessment
RAGS	Risk Assessment Guidance for Superfund
RBCA	Risk-Based Corrective Action
RfC	reference concentration

RfD	reference dose
RME	Reasonable Maximum Exposure
RSC	Relative Source Contribution
SAR	structure–activity relationship
SHEL	significant human exposure level
TDC	tolerable daily concentration
TDI	tolerable daily intake
TE	total integrated exposure
TL	tolerable level
UF	uncertainty factor

LAWS AND AGREEMENTS

CERCLA	Comprehensive Environmental Response, Compensation and Liability Act (also known as Superfund)
CEPA	Canadian Environmental Protection Act
FFDCA	Federal Food, Drug and Cosmetic Act
FIFRA	Federal Insecticide, Fungicide and Rodenticide Act
GATT	General Agreement on Tariffs and Trade
IAG	Interagency Agreement
MERA	Michigan Environmental Response Act
MOU	Memorandum of Understanding
NAFTA	North American Free Trade Agreement
RCRA	Resource Conservation and Recovery Act
SARA	Superfund Amendments and Reauthorization Act
SPS	Sanitary and Phytosanitary Agreements
TSCA	Toxic Substances Control Act

1

Introduction

MICHAEL A. KAMRIN

Institute for Environmental Toxicology, Michigan State University,
East Lansing, USA

1.1 BACKGROUND

What does harmonization mean? This is the first issue that needs to be addressed when discussing harmonization. It has variously been described as consistency, consonance, congruence, compatibility, understanding, and standardization. All of these definitions could be applied to the harmonization of risk assessment and risk management. One of the themes that runs through this book is the way that each organization responsible for risk assessment and/or risk management implicitly or explicitly defines harmonization and how this compares with the definitions used by other units that may address the same contaminants in different environmental media. Another theme is whether the same definition of harmonization should be applied to risk assessment as to risk management. A third important focus is how best to advance from the current situation to one that most appropriately harmonizes environmental risk assessment and management.

To understand why the question of harmonization of environmental risk has become an important one, it is necessary to examine the history of risk assessment and risk management for the past 50 years. During this period, a great increase in industrialization occurred followed by a corresponding large increase in concern about the impacts of the chemical, manufacturing, and power-generating industries on the environment. However, the responses of the various sectors, e.g., government, industry, public interest groups, to these concerns have not been coordinated with the result that differences have arisen with respect to the management of chemical impacts on human health due to exposures through various environmental media and also with respect to different health endpoints, particularly cancer and noncancer effects.

The fragmentation of these responses is evident in the proliferation of media- and issue-specific legislation and the issuance of numerous health

Environmental Risk Harmonization: Federal and State Approaches to Environmental Hazards in the USA.
Edited by M. A. Kamrin. Published in 1997 by John Wiley & Sons Ltd, Chichester. ISBN 0 471 97265 7

endpoint guidelines during the past few decades both internationally and in the United States. Examples of the former in the United States are the: Clean Air Act; Clean Water Act; Safe Drinking Water Act; Comprehensive Environmental Response, Compensation, and Liability Act (Superfund); Federal Insecticide, Fungicide and Rodenticide Act; and Federal Food, Drug and Cosmetic Act. Perhaps the best examples of the latter are the extensive guidelines issued by the US Environmental Protection Agency (EPA) on assessing risks from Superfund sites (US EPA 1989, 1991a,b) and guidelines on assessing specific types of toxicity, such as reproductive effects (US EPA 1996a).

This fragmentation is also evident from the way that federal and state agency responsibilities are divided. Using pesticides as an example, currently one EPA office is responsible for pesticides in raw foods, another office for the same pesticides in drinking water, a third for these pesticides in surface water, a fourth for pesticides in air, and a fifth for pesticides at hazardous waste sites. In addition, the US Food and Drug Administration (FDA) has responsibility for pesticides in processed foods and the US Department of Agriculture (USDA) has responsibility for pesticide application guidelines.

This same type of fragmentation is reflected at the state level where agencies and regulatory programs are often created in correspondence to those at the federal level. Indeed state agencies may be empowered to act as surrogates for federal agencies in carrying out certain programs. As a result, a state may have an environmental protection agency that has air, surface water, and hazardous waste units paralleling those of the EPA. However, some differences may also occur, such as a state assigning responsibility for drinking water to its public health agency instead of its environmental protection unit. In addition, even when the states have corresponding offices, risk assessment and risk management are performed under individual state environmental laws and following uniquely developed policies; laws and policies that may be quite different from those at the federal level.

Given the plethora of overlapping environmental responsibilities at both the federal and state levels coupled with media- and issue-specific legislation, it is not too surprising that a number of inconsistencies in risk management approaches and acceptable environmental exposure values have arisen. This situation in the United States has been exacerbated in recent years by the increasing interactions among the nations of the world with respect to environmental issues. The recent North American Free Trade Agreement (NAFTA) and the General Agreement on Tariffs and Trade (GATT) both have significant implications for the risk assessment and risk management practices of the signatories.

These inconsistencies have not gone unnoticed; indeed they have been the subject of a number of studies undertaken in recent years. For example, a National Academy of Public Administration (NAPA) report on the EPA (NAPA 1995) addresses the issue of inconsistencies among offices within the Agency. A draft report by the National Governors' Association (NGA) examines the diversity of state approaches to the assessment and management of risks from hazardous waste sites (NGA 1994; Appendix E). The city of Columbus, Ohio has drawn attention to inconsistencies in federal environmental regulations that address the assessment and management of risks and benefits.

The private sector has also taken note of the lack of harmonization, especially as it relates to trade issues. With the growing importance of multinational corporations and global intergovernmental cooperation, such disharmonies in environmental risk assessment and risk management can serve to inhibit collaboration and trade. This has led insurers, financial institutions, and corporations to perform their own internal risk assessments and use their economic power to push for greater uniformity.

1.2 POLICY HARMONIZATION ISSUES

To illustrate the dimensions of the harmonization issue, it is instructive to examine policy differences among agencies, specific examples of the inconsistencies that have resulted, and the implications of such differences. Considering the breadth of the issues involved, the discussion will have to be quite selective and will address only a small fraction of existing policy disparities and their implications. Other examples of these policy disparities will be evident in the chapters that follow.

A good example of disharmonies in risk management within and between federal and state agencies can be seen in the divergent policy decisions that have been made by EPA offices with respect to the acceptable risk from carcinogens. The EPA Office of Drinking Water adopted a policy that the maximum acceptable concentration for carcinogens in drinking water is zero. As zero is not measurable, this policy means that the maximum acceptable drinking water level for carcinogens will be the lowest concentration that can accurately be measured. On the other hand, the policy of the EPA Superfund Program is that a carcinogenic risk of from one additional cancer in 10 000 exposed individuals to one additional cancer in 1 million exposed individuals is acceptable. The exact acceptable risk value at a Superfund site will thus depend on a number of factors other than the absolute cancer risk value.

The FDA has not established an explicit acceptable risk level for carcinogens in food and, under some legislation, incorporates factors other than cancer risk in arriving at acceptable maximum levels in foods. In contrast, the EPA has established a maximum target cancer risk of one additional cancer in 1 million exposed individuals for setting tolerances for pesticides in raw foods. Recent legislation, the Food Quality Protection Act of 1996, includes language extending this policy to maximum allowable concentrations of pesticides in processed foods.

States may use their own criteria, criteria that may differ from the one chosen by a corresponding federal agency, in setting acceptable risk levels for carcinogens. For example, the state of Michigan recently changed its criterion from one additional cancer in 1 million exposed individuals to one additional cancer in 100 000 exposed individuals for determining maximum acceptable environmental concentrations at hazardous waste sites. The acceptable cancer risk values vary from state to state.

In addition to the risk management policy differences detailed here, there are risk assessment policy differences that also can have a significant impact on the determination of acceptable environmental exposure levels for carcinogens. For example, there is the issue of whether to extrapolate from rodents to humans on the basis of skin surface area or body weight. Although a tentative compromise has been agreed upon, the official EPA position corresponds to the former approach and the FDA position to the latter (US EPA 1992). This one difference can lead to a 10-fold difference in the acceptable environmental exposure limits.

Another difference in cancer risk assessment, and one that is seen very clearly at the international as well as the national level, is the appropriate technique for extrapolating from high-dose experimental data to the much lower dose levels corresponding to the acceptable risk values, e.g., one in a million. This is especially problematic for carcinogens that show a nonlinear dose response. In cases such as these, some government entities utilize a safety factor approach as opposed to a mathematical extrapolation. Combined with differences in the assumptions used in extrapolating from experimental animals to humans, this has resulted in orders of magnitude differences in cancer risk values and thus in acceptable environmental concentrations (Table 1.1).

Increased attention to these and other inconsistencies with respect to cancer and noncancer risk assessment have led to the convening of a number of expert groups and the publication of a number of extensive reports. One example is the National Academy of Science volume entitled *Science and Judgement in Risk Assessment* (NRC 1994). More recently, the EPA has issued a

Table 1.1 Variability in calculated lifetime dioxin dose to cause one additional cancer in 1 000 000 exposed individuals (EPA 1988)

Agency	Dose (pg/kg per day)*
US Environmental Protection Agency	0.006
State of California	0.007
US Centers for Disease Control	0.03
US Food and Drug Administration	0.06
Germany	1.0
State of New York	2.0
The Netherlands	4.0
Canada Health and Welfare	10.0

*picograms of dioxin per kilogram of body weight per day.

draft report containing proposals for significant revisions in its cancer risk guidelines (US EPA 1996b) and the Commission on Risk Assessment and Risk Management (CRARM 1996; Appendix B), mandated under the Clean Air Act Amendments of 1990, has issued a draft report containing a number of harmonization-related recommendations.

1.3 EXAMPLES OF HARMONIZATION ISSUES

Table 1.1 in the previous section illustrated the differences in carcinogenic risk values for dioxin generated by a variety of governmental units in the United States and abroad. It illustrates the impact that policy choices can have on risk assessment as well as risk management. As such, it also reflects how far the current situation diverges from the ideal separation of risk assessment and risk management, which is suggested in the National Academy of Sciences landmark report *Risk Assessment in the Federal Government: Managing the Process* (NRC 1983).

Another situation that reflects the impact of political considerations on risk assessment and risk management is the difference in the treatment of artificial sweeteners in the United States and Canada. At present, the US government policy is that cyclamates are forbidden to be used as an artificial sweetener while saccharin is permitted to be used. The Canadian government has taken the exact opposite position and currently bans saccharin use while permitting the use of cyclamates. There is little dispute about the scientific evidence concerning the risk from these two types of sweeteners so this difference can only be accounted for on policy grounds.

A good illustration of the interactions among federal agencies and between federal and state agencies occurred in Michigan a few years ago. Under FDA regulations, fish sold in interstate commerce can contain no more than 0.3 parts per million (p.p.m.) of the pesticide dieldrin. A group in Michigan, to be environmentally protective, proposed to take fish wastes left after Great Lakes fish are cleaned and turn them into compost. This was in lieu of sending approximately 6–7 million pounds of fish wastes to solid waste landfills.

This group demonstrated that the fish composted very effectively and over a short period of time when mixed with sawdust, and it was proposed that the compost be sold to gardeners to cover the costs of the composting operation. However, the Michigan Department of Natural Resources (MDNR), the agency responsible for hazardous waste site cleanup, had the compost analyzed and found it to contain 80 parts per billion (p.p.b.) of dieldrin, a value that exceeded the Michigan soil clean-up criterion for this pesticide at that time. Thus, the state ruled that this compost could not be applied to soil.

The discrepancy between the management of dieldrin in fish versus soil is even greater than the maximum exposure numbers might indicate. The fish exposure to dieldrin is by ingestion and the EPA default average daily value for fish consumption is 6.5 g; the soil exposure is mainly by ingestion and the EPA default for consumption of soil is 0.1 g (for the most highly exposed subpopulation, children). A simple calculation shows that, using the default values, exposure to fish would result in 65 times higher daily doses than exposure to soil. This discrepancy is quite large and, contrary to what might be expected in managing risks to human health, exposure from the larger source was allowed while that from the smaller one was forbidden.

While this might be seen as an example of Federal and State disharmony, it is more complex than this. The Michigan risk management approach for toxic chemicals at hazardous waste sites reflects EPA policies for addressing similar situations so what appears on the surface to be a state/federal disparity is, at its roots, a reflection of the difference between EPA policy addressing chemicals at hazardous waste sites and FDA policy towards chemicals in fish sold in interstate commerce. This situation also has state/federal dimensions as Michigan hazardous waste policies are somewhat different from those of the EPA Superfund Program.

This dieldrin issue can be taken one step further by comparing the acceptable soil levels promulgated at that time by the MDNR Environmental Response Division to the maximum acceptable ambient air level promulgated by the MDNR Air Quality Division. This latter value of 0.0002 $\mu g/m^3$ in air for dieldrin corresponds, using standard inhalation exposure assumptions, to an acceptable daily intake of about 1.4×10^{-4} mg. When a similar calculation is performed using the soil criterion value, the maximum acceptable daily intake is about 6×10^{-6} mg. This is a difference of between one and two orders of magnitude.

As in the previous example comparing acceptable levels in fish and compost, the situation for soil versus air exposures is more complex than it appears on the surface. On its face, this discrepancy appears to reflect disharmonies among state environmental units, in this case in the same department, the MDNR. However, it also reflects differences at the federal level. The Superfund Program has a large number of guidance documents which are used by the states to set criteria values in soil at hazardous waste sites. The federal government does not have a corresponding set of documents for ambient air. In the absence of specific guidance, the states have developed a variety of approaches in establishing acceptable ambient air exposure limits. The result is that the acceptable values in one state are often not congruent with those from neighboring states as well as with exposures allowed in other media by other government units in the same state.

Another example of a risk management inconsistency that also occurred a few years ago involved an attempt by the US Corps of Engineers to land-apply dredged sediments containing 10–15 p.p.m. arsenic. This was not allowed by the MDNR as these concentrations exceeded the existing hazardous waste site soil clean-up criterion of 5.6 p.p.m. The amount of resources required to deposit these sediments in hazardous waste landfills is very great and the result was that the sediments were left in place.

However, other EPA regulations permit the application of sewage sludges with higher levels of arsenic, up to 41 p.p.m., to agricultural soils. This inconsistency obviously has significant implications if large amounts of resources are being used to reduce levels in sediments while much higher arsenic levels are not only allowed but are created at other sites.

As in the previous situations, an apparent Federal and State inconsistency actually reflects a federal/federal disharmony as the EPA office dealing with solid wastes, e.g., sewage sludge, does not follow the same risk assessment and risk management policies as the office that manages hazardous wastes; although the chemicals involved are the same in many instances. In this case the MDNR Division corresponding to the EPA Superfund Program office followed this office's guidance and thus came into conflict with another unit within the EPA.

While a number of changes have occurred in Michigan since these events happened, particularly the establishment of a Department of Environmental Quality separate from the DNR and revision of the laws governing hazardous waste sites, this has not resulted in elimination of inconsistencies (see Chapter 10, Part II). Instead, in most cases, these new laws have only resulted in quantitatively different disparities.

1.4 IMPLICATIONS OF DISHARMONIES

It could be argued that many of the differences in the way that risk is assessed and managed are not really discrepancies; instead they may reflect real differences in circumstances that warrant different approaches. This has been the argument used to justify the very great differences in allowable exposures in the workplace and the environment. In this case, it has been argued that workers are a special class of the general population that is made up of individuals who are neither very young nor very old and are likely to be generally healthy individuals. Thus, the conservative assumptions designed into risk assessment and risk management to protect the most sensitive populations do not apply to them.

It is often further argued that, in contrast to the environmental situation, these individuals have the use of personal protective equipment to reduce exposure and that the percentage of time spent at work is much less than that spent in the environment. A last justification that is sometimes expressed is that the workers' exposures are voluntary and that they are aware of the risks and can weigh these risks against the benefits in deciding whether or not to work at a particular job.

In recent years, these justifications have come under increasing scrutiny and there appears to be a movement towards making the occupational standards more stringent and more consistent with the environmental ones. A possible contributor to this change is the recognition that occupational exposures are generally much higher than environmental ones and that workers have exhibited clinically significant effects in some cases, e.g., asbestos and benzene. Perhaps, the publicity surrounding the US Supreme Court decision about exposure of females to lead in the workplace contributed to this recognition (*New York Times* 1991). Also, in an era where cost–benefit has become more important, the expenditure of large sums to reduce already very low environmental exposures while ignoring much higher occupational ones appears less and less defensible.

At the same time as occupational/environmental comparisons gained in importance, comparisons among environmental risks also have become more popular, both at the federal and state level. A number of state comparative risk reports attest to this popularity (MDNR 1992; California EPA 1994). These comparisons are driven by the same cost–benefit considerations and, in some cases, have highlighted discrepancies between and among agencies and offices. In addition, society is asking more difficult questions about the results of risk assessment and risk management, such as the implications of current policies for environmental justice.

In such circumstances, the policy discrepancies such as those outlined and illustrated in this chapter are seen in a less sanguine light. This is especially true as resources for environmental protection are reduced and priorities must be established among competing risks in different environmental media. Thus, it is to be expected that risk managers will be under increasing pressure to make decisions under a consistent set of guidelines for both assessment and management of risk.

1.5 CONCLUSIONS

The aim of this book is to provide a synopsis of the current status of Federal and State risk harmonization in the United States and to provide some guidance as to inconsistencies that should be reduced, as well as the route to take to reach this goal. The parts of this picture are represented by the descriptions given by international, federal, and state agency scientists of their current approaches to risk assessment and risk management. The reader can appreciate the degree and character of the disparities among these various agencies through a careful reading of each of these chapters.

In addition, the chapter summarizing the workshop discussions points up the existence of serious divisions that are not being addressed at present as well as areas of current consensus and areas where agreement might be readily achieved. In addition, the workshop participants provide a glimpse of what the future may bring; which approaches to greater consistency are most likely to be successful and how general trends, such as devolution of regulatory powers to the states, might impact on the success of these approaches.

REFERENCES

California Environmental Protection Agency (1994). Toward the 21st Century: Planning for the Protection of California's Environment. California Comparative Risk Project Final Report. Sacramento, CA.

Commission on Risk Assessment and Risk Management (1996). Risk Assessment and Risk Management in Regulatory Decision-making (Draft). Washington, DC.

Michigan Department of Natural Resources (1992). *Michigan's Environment and Relative Risk*. Lansing, MI.

National Academy of Public Administration (1995). *Setting Priorities, Getting Results: A New Direction for EPA*. Washington, DC.

National Governors' Association (1994). Risk in Environmental Decisionmaking: A State Perspective. Working Paper. Denver, CO.

National Research Council (1983). *Risk Assessment in the Federal Government: Managing the Process*. National Academy Press, Washington, DC.

National Research Council (1994). *Science and Judgment in Risk Assessment*. National Academy Press, Washington, DC.

New York Times (1991). Court backs right of women to jobs with health risks. March 21, p A1, New York.

US Environmental Protection Agency (1988). Draft Report: A Cancer Risk-specific Dose Estimate for 2,3,7,8-TCDD. EPA/600/6–88/007Aa. Washington, DC.

US Environmental Protection Agency (1989). *Risk Assessment Guidance for Superfund*, Vol. 1. *Human Health Evaluation Manual* (Part A). Interim Final. EPA/540/1–89/002. Office of Emergency and Remedial Response, Washington, DC.

US Environmental Protection Agency (1991a). *Risk Assessment Guidance for Superfund*, Vol. 1. Part B. EPA 9285.7–01B. Office of Emergency and Remedial Response, Washington, DC.

US Environmental Protection Agency (1991b). *Human Health Evaluation Manual, Supplemental Guidance: Standard Default Exposure Factors*. OSWER Directive 9285.6–03. Office of Solid Waste and Emergency Response, Washington, DC.

US Environmental Protection Agency (1992). Draft report: a cross-species scaling factor for carcinogen risk assessment based on equivalence of $mg/kg^{3/4}/day$. *Federal Register* **57**: 24152–24173.

US Environmental Protection Agency (1996a). Guidelines for reproductive toxicity risk assessment. *Federal Register* **61**: 56274–56322.

US Environmental Protection Agency (1996b). *Proposed Guidelines for Carcinogen Risk Assessment*. EPA/600/P-62/003c. Washington, DC.

2

Harmonization of Approaches to the Assessment of Risk from Exposure to Chemicals

CYNTHIA SONICH-MULLIN
International Programme on Chemical Safety, World Health Organization,
Geneva, Switzerland

2.1 INTRODUCTION

The International Programme on Chemical Safety (IPCS) has embarked on a large-scale project on the harmonization of approaches to the assessment of risk from exposure to chemicals. The IPCS recognizes that the success of this project lies in the commitment and active participation of a number of partners, both nationally and internationally, and that its role is not only to contribute to the resolution of the scientific issues raised, but perhaps more importantly, to identify partners and to coordinate activities among these partners.

The first part of this paper will describe the project's scope and objectives as they relate to the overall efforts of the IPCS. This will be followed by a summary of completed and ongoing efforts, focusing on the progress made to date. Current efforts have suggested additional activities which will be summarized in the plan of action. The last section will provide a brief discussion of the outlook for the future.

2.2 BACKGROUND

The IPCS is a joint program of the World Health Organization (WHO), the International Labor Organization (ILO), and the United Nations Environment Program (UNEP). There are two major objectives of the IPCS.

Cynthia Sonich-Mullin is now located at the US Environmental Protection Agency, Cincinnati, Ohio.

1. *To establish the scientific basis for assessment of the risk to human health and the environment from exposure to chemicals through international peer-review processes, as a prerequisite for the promotion of chemical safety.* This involves the development and validation of specific methodologies, as well as the evaluation of risks associated with exposure to specific chemicals.
2. *To provide technical assistance in strengthening national capacities for the sound management of chemicals.* Methodologies and guidance values alone cannot ensure safety if they are not understood and properly implemented.

IPCS activities can be categorized into six areas (Table 2.1). These include the evaluation of risks on both chemical-specific and class-specific bases, and the development and continual improvement of risk assessment methodologies. While these two IPCS activities are those better known among risk assessors, IPCS efforts also address the prevention of chemical emergencies and poisonings, as well as the management of chemical emergencies and the treatment of poisoning patients. In addition, IPCS devotes substantial efforts to technical cooperation, including training and human resources development, as the value of the methods developed and the usefulness of the evaluation of risks lie in the clear understanding and appropriate implementation of this information. Through each of these activities, the need for harmonization has clearly emerged and continues to be identified as a priority.

Broad recognition of the international significance of risk harmonization can be traced to the discussions at the United Nations Conference on the Environment and Development (UNCED), held in Rio de Janeiro in 1992. At this conference, and as stated in Agenda 21, Chapter 19, UN Member States recommended that IPCS be the nucleus for international cooperation on environmentally sound management of toxic chemicals. This was subsequently discussed by over 114 governments at the 1st International Forum on Chemical Safety (IFCS) held in 1994. The IFCS (or Forum) was established to provide policy guidance and develop coordinated and integrated international strategies to implement the goals set out in Agenda 21. The Forum is a mechanism for increased cooperation between governments and other parties involved in

Table 2.1 Activities of the International Programme on Chemical Safety

---

Risk evaluation of priority chemicals
Development of risk assessment/management methodologies
Prevention and management of chemical emergencies
Prevention and treatment of poisonings
Promotion of technical cooperation
Promotion of training

---

strengthening chemical safety. The governments attending the first Forum meeting identified harmonization as a high priority area and endorsed the UNCED recommendations that IPCS serve as the nucleus of the efforts in this area.

Following the recommendations of UNCED in 1992, the IPCS harmonization project was officially initiated at a planning meeting in the fall of 1993 (IPCS 1993). This meeting brought together individuals with diverse scientific expertise and experiences from a number of countries, as well as representatives of both national and international organizations, including nongovernmental organizations. The discussions held at the planning meeting resulted in the endorsement, support, and commitment of those present for this ambitious effort. Another important result was the agreement reached on the definition of harmonization: an understanding of the methods and practices used by various countries and organizations to develop confidence in and acceptance of assessments using different approaches.

2.3   WHAT IS HARMONIZATION?

To understand better what harmonization is, it is helpful, first, to describe what it is not. Harmonization, in the risk context, should not be equated with standardization. It is not a goal of the IPCS project to standardize risk assessments globally, as that would be neither appropriate nor feasible. Instead, harmonization can be thought of as an effort to strive for understanding and consistency among approaches.

This goal can be accomplished, hopefully, in a step-wise fashion (Table 2.2). The first step is to understand the methods used by individual organizations, through increased communication. It is hoped that this will increase confidence in and acceptance of assessments using different approaches, and result in a willingness of all to work toward convergence of such approaches as a long-term goal. However, it should again be emphasized that this convergence is not the same as a standardization of approaches.

Table 2.2   What is harmonization?

Understanding the methods and practices used by various countries and organizations
Developing confidence in and acceptance of assessments using different approaches
Willingness to work toward a convergence of methodologies as a long-term goal
Harmonization is not standardization

Table 2.3 Advantages of harmonization

Provides framework for comparing information
Advances understanding the basis of exposure standards
Enables progress toward common classification and labeling schemes
Leads to savings of time and expense through information sharing
Promotes credible science

2.4 WHY HARMONIZE?

Why is this harmonization effort important? There are a number of reasons (Table 2.3). First, this project will produce a framework for a comprehensive comparison of information and approaches. From such comparisons, a better understanding of the bases for exposure standards used by different countries and organizations can be achieved. This increased understanding can contribute to the development of common classification and labeling schemes; a process now being carried out under the IOMC[1] Coordinating Group for the Harmonization of Chemical Classification Systems consisting of the ILO, Organization for Economic Cooperation and Development (OECD), and United Nations Committee of Experts on Transport of Dangerous Goods (UNCETDG). The sharing of information can also result in a significant savings of time and expense, a benefit which should not be underestimated in this time of limited resources and many needs. Finally, it is expected that the use of credible science will increase through better communication among organizations and peer review of assessments and assessment procedures.

2.5 PLAN OF ACTION

A report summarizing the conclusions and consensus of those present at the 1993 Planning Meeting included an endorsement of the project, a working definition of harmonization, and an initial action plan. As part of this action plan, a number of priority areas were identified (Table 2.4) and it was decided that the project be undertaken on an endpoint-specific basis. The specific priority areas were chosen on the basis of a number of criteria including:

---

[1] The IOMC, or Inter-Organization Programme for the Sound Management of Chemicals, is a program designed to be a cooperative undertaking among six intergovernmental organizations: UNEP, ILO, Food and Agriculture Organization (FAO), WHO, United Nations Industrial Development Organization (UNIDO), and OECD.

1. potential public health impact;
2. availability of both national and international guidance documents;
3. probability of short-term success;
4. lack of consideration of the endpoint in any other fora;
5. existence of at least partial agreement on data requirements and study protocols.

Table 2.4 Priorities for the International Programme on Chemical Safety harmonization action plan

First priority
    reproductive/developmental toxicity
    carcinogenicity
    mutagenicity
Second priority
    neurotoxicity
    general systemic toxicity
    immunotoxicity

All endpoints identified were considered to be of high priority; however, given the limitation of available resources, one group of endpoints was designated as a first priority and another group as a second priority. Reproductive and developmental toxicity was identified as the endpoint most ripe for harmonization because of the potential for success in the short term, and was, therefore, given the highest priority. Carcinogenicity was also included as a first priority because the assessment of carcinogenic risk involves complex scientific questions; it was the opinion of the Planning Meeting participants that success in addressing this endpoint required that efforts begin as soon as possible.

Once priorities were established, the next step in the action plan was to gather information about risk assessment and management approaches currently in use. This activity is ongoing (Table 2.5) and includes efforts to develop links with various organizations, some new to IPCS, to both enhance and enrich the process, and to avoid duplication of effort. As information is

Table 2.5 Ongoing harmonization activities

Development of links with various organizations to:
    enhance and enrich the process
    avoid duplication of effort
Collection and compilation of all available guidance
documents and other relevant information

collected, IPCS and OECD is jointly developing an inventory of risk assessment methodologies for widespread, multimedia circulation.

While the collection and organization of information are quite simple, they meet an important need of the international community. The importance of such simple steps to harmonization suggests that the most successful strategy may be to focus first on common sense approaches rather than the more complex scientific issues. These relatively simple but important efforts can then serve as the bases for further work toward harmonization.

In light of the far-reaching goals of this project, it was essential for IPCS to define some bounds. One such decision was to focus on human health rather than environmental risk assessment, although it is understood that this distinction is not always possible, nor plausible. Given this initial focus, the project includes consideration of all aspects of the National Academy of Sciences (NAS) risk assessment paradigm: hazard identification, dose–response assessment, exposure assessment, and risk characterization (NRC 1983).

2.6 PROGRESS

Once the action plan and priorities were established, IPCS convened a series of workshops, the first of which focused on reproductive and developmental toxicity. This workshop was developed jointly with OECD which, through its pesticides pilot project, had identified specific needs in this area. It was quickly discovered that although this area was selected as ripe for harmonization, resolving the issues that hinder harmonization of methodologies would not be easy. The workshop participants did, however, make many recommendations for future efforts; one of particular importance was to harmonize terminology used in this area. This resulted from the realization that the definitions used by different organizations for relatively simple terms such as 'reproductive toxicity' or 'developmental toxicity' are quite different. Another recommendation was to harmonize the definition of 'adverse effect' as used in risk assessment. Further, the need for efforts to achieve a common reporting format for assessments was clearly identified.

Following this workshop, a scoping meeting was convened to discuss issues related to the harmonization of approaches to carcinogenicity and mutagenicity, both separately and together. The meeting participants provided a number of recommendations for future efforts. One was to conduct chemical-specific comparisons as the bases for understanding the differences in conducting cancer risk assessments in different countries. Participants concluded that this was necessary because guidelines alone do not provide a clear view of how assessments are actually conducted for chemicals of concern. It was suggested that five or six chemicals be carefully chosen to illustrate how different countries or organizations implement their guidelines. The chemicals chosen should provide the opportunity to address the many different kinds of data used in the conduct of risk assessments on chemicals with carcinogenic potential.

In the area of mutagenicity, the workshop participants decided to define this effect as the production of germ cell changes. They also indicated that a scheme for the quantification of mutagenic assessments was necessary. A subgroup on mutagenicity moved forward to develop such a scheme and its recommendations were published in the journal, *Mutation Research*, in 1996 (Ashby *et al.* 1996). Journal publication was chosen as the mechanism by which comments on the proposed scheme could best be obtained. Once comments are received, the quantitative scheme will be refined.

An additional recommendation of the 1993 Planning Meeting was that, given the magnitude of the project, a Steering Committee be formed. This was accomplished in parallel with the convening of the scientific workshops. The purpose of the Steering Committee is to provide guidance and overall direction to the project and, in addition, to provide guidance on and work toward implementation of the scientific recommendations made by the work groups, i.e., to put into practice the outputs of the projects. To accomplish these goals, the Steering Committee is composed of individuals with expertise in risk assessment and scientific fields, but more importantly, includes individuals who have a strong influence in their countries or organizations and thus can provide guidance that has a good chance of being translated into reality. The best scientific decisions will be of little significance if they cannot be implemented.

The first meeting of the Steering Committee was convened in June 1995 to review the progress made at the first two workshops. It was the general consensus of this body that the process was working, and that the project was headed in the right direction. The committee also identified several areas of future work, including environmental risk assessment methods, methods for assessing risk of chemical mixtures, the use of epidemiology data in risk assessment, and integrated risk assessment methodologies.

As a result of the recommendations of the Steering Committee and their endorsement of the work that was done on reproductive and developmental toxicity, a second meeting on this endpoint was convened. Outputs of this meeting included a consensus on the work necessary to harmonize terminology for this particular endpoint. Efforts of national teratology societies and the International Federation of Teratology Societies were defined further. Recommendations were made to enhance this work to provide the information necessary to lead to a full harmonization of the use of such terminology. The role of the International Federation of Teratology Societies in taking the lead on this issue cannot be underestimated, and this will hopefully serve as a model for other scientific societies and organizations that have a contribution to make to the terminology effort as well as to other project activities. It is only through discussions at the scientific level, and the circulation and use of these definitions among scientists, that a common terminology in specific scientific areas will become a reality.

Much discussion revolved around the use of the term 'adversity' at this second workshop and this issue illustrated the difficulties involved in achieving harmonization. While the first workshop eventually agreed on a description of 'adversity', it didn't take long for this second workshop of scientists (some of whom participated in the first workshop) to come to a conclusion that they disliked the definition agreed to by the first workshop. While this is part of the process, it provides one indication of the magnitude of the work ahead.

2.7 FUTURE EFFORTS

It is clear that future efforts will cover a wide range of activities. In the area of carcinogenicity, the process of chemical-specific comparisons will begin soon. The first step will be a meeting to select the chemicals to be assessed. This will be followed by performance of the actual assessments of the chemicals selected by specific countries or organizations, and a peer review meeting to compare the resulting assessments. In this way, existing guidelines and the way they are implemented can be compared and opportunities for harmonization identified. Complementing this effort is an Australian Department of Health survey to determine which countries have developed their own carcinogen guidelines and which use those of others and further, to review existing guidelines that may be applicable to Australia. Through both of these efforts the issues related to harmonization of approaches to the assessment of carcinogenic risk will be identified and addressed, and additional areas of research will be delineated.

In the area of mutagenicity, both the content and implementation of existing guidelines will be carefully examined. Previous work has shown that, in many cases, the written words reflect legally-binding mandates; however, the implementation is more reflective of scientific judgements. As a result, approaches that may appear diverse on paper may not be that different in practice. Further, existing mutagen classification schemes will be compared with the goal of developing international guidelines for the quantitative assessment of mutagenic risk.

There are a number of additional efforts planned (Table 2.6). One is conti-

Table 2.6 Planned efforts

Address endpoint-specific issues
Advance consistency in terminology
Examine role of epidemiology in risk assessment
Address exposure assessment issues
Highlight the important aspects of reporting risk assessments
Include environmental risk assessment issues

nuing work with the OECD on the harmonization of risk assessment terminology. This joint project emanated from the IPCS harmonization project as well as related efforts in both IPCS and OECD. It is being conducted under the framework of the IOMC.

IPCS will also begin to identify and elucidate further the role of epidemiology in the risk assessment process. This is one issue that crosses all approaches and has been identified as a priority area in all the workshops thus far convened. A workshop on this topic will be held in conjunction with a major epidemiology society meeting.

Additional areas that will be addressed are neurotoxicity and immunotoxicity endpoints, as well as approaches to assessing general noncancer effects, such as use of the benchmark dose and significance of body weight gain as an indicator of toxicity. The neurotoxicity discussions may be viewed as premature given that such guidelines are currently under development in many countries. However, early involvement in that process might produce better harmonization efforts in this area.

Further, guidelines for exposure assessment will be reviewed to identify issues hindering harmonization of such methodologies. Issues related to the reporting of risk assessments will be also explored. Last, the scope of the overall project will be expanded to include environmental risk assessment methodologies, an effort that will be undertaken cooperatively with OECD.

2.8 FUTURE ISSUES

While it is clear that there are many advantages of harmonizing approaches to risk assessment and that this is a goal shared by all partners, there are serious obstacles to overcome. One is the willingness of scientists and regulators to make compromises in the interest of harmonization. It may not be easy for individuals with diverse backgrounds (and comfort with long-standing 'traditional' methods or approaches), to resolve issues pertaining to specific risk assessment approaches. This can be due to a variety of considerations including socio-economic and political systems, language, and scientific discipline. However, once such compromises are made, the strength of consensus should become evident. If countries or organizations can clearly understand the benefits that such efforts can achieve, then harmonization is a realistic goal.

The IPCS efforts so far clearly show that, to be successful, harmonization cannot (and should not) be dictated (Table 2.7). Approaches to risk assessment intuitively differ by discipline and mandate. Countries and organizations, over time, have developed cultural or historical ways of viewing risk and dealing with specific risk issues (e.g., differences in the way that food risks and chemical risks are viewed). However, harmonization of approaches can result from open and continuing discussions among members of the scientific community, including exchange of information and mutual understanding of

Table 2.7 Future considerations

Harmonization cannot be dictated
Harmonization will result from:
   scientific discussions
   information exchange
   mutual understanding of goals and objectives

goals, objectives, and rationale. Such interactions can also result in the best use of all available data and information, while recognizing and acknowledging differences in approaches.

In light of this, the IPCS project uses a collective and collaborative 'bottom-up' approach. The IPCS serves as the coordinating institution to facilitate this effort and hopes that other organizations/institutions/societies will each assume the responsibility for a specific activity or task. This 'bottom-up' approach begins with information exchange and discussions at the scientific level followed by the development of recommendations that include consideration of the issues involved in implementation. To achieve such recommendations, all parties (scientists from all sectors, regulators, industry, and the public) have been willing to work collectively and cooperatively. This effort can continue and expand if it is recognized that the goal is not to harmonize the past, but to move forward in a participatory manner toward improved risk assessment, and create the future together.

REFERENCES

Ashby J, Waters MD, Preston J, Adler ID, Douglas GR, Fielder R, Shelby MD, Anderson D, Sofuni T, Gopalan HNB, Becking G, Sonich-Mullin C (1996). IPCS harmonization of methods for the prediction and quantification of human carcinogenic/mutagenic hazard, and for indicating the probable mechanism of action of carcinogens. *Mutation Research* **352**: 153–157.

International Programme on Chemical Safety (1993). IPCS Planning Meeting on Harmonization of Approaches to the Assessment of Risk from Exposure to Chemicals. Ottawa, Canada, 30 August–1 September 1993. IPCS/PAC/93.11. World Health Organization, Geneva.

National Research Council (1983) *Risk Assessment in the Federal Government: Managing the Process.* National Academy Press, Washington, DC.

3

Approach to Risk Assessment for Priority Substances in Canada: Novel Aspects

BETTE MEEK AND KATHY HUGHES

Environmental Health Directorate, Health Canada, Ottawa, Ontario, Canada

3.1 INTRODUCTION

The Canadian Environmental Protection Act (CEPA), authorizes the Ministers of the Environment and of Health to investigate a wide variety of substances, that may be present in the environment and cause adverse effects on the environment or on human health. The Ministers (Environment and Health) must establish a List of Priority Substances (PSL). The first of these lists (PSL 1) was released in February, 1989 and comprised 44 substances, including discrete chemicals, classes of substances and complex mixtures (Table 3.1), for which assessment reports were published within a legislated 5-year time frame (by February, 1994). The second PSL (PSL 2) was released in December, 1995 and contains 25 substances (Table 3.2) which must also be evaluated within a 5-year time frame by December, 2000.

These lists were developed by multistakeholder advisory committees to the Ministers of Environment and Health. To assist in this process, for the second PSL, information on presence in the Canadian environment and toxicity was systematically screened for over 600 candidate substances (Government of Canada 1996).

As prescribed within CEPA, the intent of the evaluations of Priority Substances is to assess whether or not these substances enter or are present in the environment in concentrations:

1. having or that may have an immediate or long-term harmful effect on the environment;

This chapter is a Canadian Government work and, as such, is in the public domain.

Table 3.1   First Priority Substances List Canadian Environmental Protection Act

Aniline	Dioxins
Arsenic and its compounds	Furans
Benzene	Hexachlorobenzene
Benzidine	Inorganic fluorides
Bis(2-chloroethyl) ether	Man-made vitreous fibers
Bis(chloromethyl) ether	Methyl methacrylate
Bis(2-ethylhexyl) phthalate	Methyl tertiary butyl ether
Cadmium and its compounds	Nickel and its compounds
Chlorinated paraffin waxes	Non-pesticidal organotins
Chlorinated wastewaters	Polycyclic aromatic hydrocarbons
Chlorobenzene	Pentachlorobenzene
Chloromethyl methyl ether	Pulp mill effluents
Chromium and its compounds	Styrene
Creosote-impregnated waste materials	Tetrachlorobenzenes
Dibutyl phthalate	1,1,2,2-Tetrachloroethane
3,3'-Dichlorobenzidine	Tetrachloroethylene
1,2-Dichloroethane	Trichlorobenzenes
1,2-Dichlorobenzene	1,1,1-Trichloroethane
1,4-Dichlorobenzene	Trichloroethylene
Dichloromethane	Toluene
3,5-Dimethylaniline	Waste crankcase oils
Di-n-octyl phthalate	Xylenes

Table 3.2   Second Priority Substances List Canadian Environmental Protection Act

Acetaldehyde
Acrolein
Acrylonitrile
Aluminum chloride, aluminum nitrate, aluminum sulfate
Ammonia in the aquatic environment
1,3-Butadiene
Butylbenzylphthalate
Carbon disulfide
Chloramines
Chloroform
N,N-Dimethylformamide
Ethylene glycol
Ethylene oxide
Formaldehyde
Hexachlorobutadiene
2-Methoxy ethanol, 2-ethoxy ethanol, 2-butoxy ethanol
N-Nitrosodimethylamine
Nonylphenol and its ethoxylates
Phenol
Releases from primary and secondary copper smelters and copper refineries
Releases from primary and secondary zinc smelters and zinc refineries
Releases of radionuclides from nuclear facilities (impacts on nonhuman species)
Respirable particulate matter less than or equal to 10 μm
Road salts
Textile mill effluents

2. constituting or that may constitute a danger to the environment on which human life depends; or

3. constituting or that may constitute a danger in Canada to human life or health.

The assessments of Priority Substances under CEPA correspond to the risk assessment phase of a risk assessment/risk management paradigm. Aspects of risk management are not considered at this stage. Rather, the assessments form the basis upon which the priorities for options to control risks to human health or to the environment are judged. Risk management is addressed in a separate, subsequent stage entitled the strategic options process, which involves consultation with affected or interested parties. In this phase, recommendations concerning the need for, and development of, control strategies are made following a judicious balancing of the estimated risks against the associated costs and feasibility of controls, and/or benefits to society.

Initial steps in the preparation of the health assessments include compilation of drafts and review by senior technical staff of the Environmental Health Directorate of Health Canada as well as external technical peer review by experts internationally. This latter phase entails an independent review by individual experts and/or where considered warranted, consensus review by a Task Group of experts. In subsequent stages, assessments are approved by an internal committee of Health Canada composed of representatives of the Environmental Health and Food Directorates and an interdepartmental management committee of Health and Environment Canada. For substances on the second PSL, there will also be a public review period in the late stages of completion of the assessments. In addition to assessment reports which have been published for each of the substances included on the first PSL, a compilation of tolerable intakes/concentrations and tumorigenic doses/concentrations has also been released to provide assistance in the development of guidelines for the quality of various media (Health Canada 1996).

In this chapter, the general principles developed for the assessment of risk to human health for Priority Substances under CEPA are addressed with special reference to unique and evolving aspects.

3.2 ASSESSMENT OF POPULATION EXPOSURE

Exposure to environmental substances may occur by inhalation, ingestion and/ or dermal absorption from air, water, food, soil, and through the use of consumer products. Estimation of total exposure (intake) from all sources is critical in assessing the overall magnitude of risk. This 'multimedia' approach also sets the stage for any subsequent development of measures which are most effective for human health protection by identifying the relative magnitude of the contribution of each pathway to total exposure.

Standardized reference values for body weights, the volume of air breathed and quantities of food, water and soil ingested and, to the extent possible, information on activity patterns of the exposed population(s), form an integral part of the estimation of exposure from all sources for Priority Substances. Reference values for these parameters have been developed for five discrete age groups with varying potential for exposure to environmental substances: infants, pre-school children, elementary school children, teenagers, and adults. Where possible, these reference values have been developed on the basis of national surveys conducted by Health Canada, including those on fitness and consumption of drinking water and food (Nutrition Canada 1977; Environmental Health Directorate 1981; Health Canada 1994).

It should be emphasized that the reference values and ranges of mean concentrations in environmental media or food from national surveys on which estimates of exposure are based, are representative for average members of the general population of Canada. However, exposure to chemical substances for some segments of the population may be greater than that for the population at large and the exposure assessment takes this into account. 'High-exposure subgroups' are selected on a case-by-case basis. In addition to the magnitude of the estimated exposure for various subgroups, the size of the exposed population and availability of reliable data on concentrations in relevant media as well as consumption are also considered.

3.3 RECENT DEVELOPMENTS: EXPOSURE ASSESSMENT

The approach to estimation of exposure for Priority Substances is evolving, to reflect recent developments in this area and to provide information better tailored to suit the needs of the subsequent but separate risk management phase. Indeed, the approach to exposure assessment for PSL 2 substances will likely be hierarchical, with the extent of characterization being inversely proportional to the size of the margin between estimates of environmental levels and doses or concentrations which induce critical toxic effects. In addition to deterministic estimates, probabilistic estimates based on uncertainty analysis may be developed. Where data permit, a range of exposure scenarios will be presented for the general population and for several of the more highly exposed subgroups (particularly those in the vicinity of industrial sources as these are most amenable to control under CEPA).

The data on exposure available for most of the substances included on PSL 2 are far more limited than that for PSL 1 compounds. This will necessitate greater reliance on modeling (e.g., fugacity and consumer exposure), where it is considered to contribute reliably. To some extent, the multimedia estimates of exposure for PSL 1 compounds lend support to the use of fugacity modeling, based predominantly on information on physical/chemical properties, as a basis for prediction of, at least, principal routes of exposure. Multimedia exposure estimates for PSL 1 compounds based on monitored concentrations confirmed that for volatile organic compounds, air was the principal medium of exposure; for less volatile compounds, food was the most important source of exposure. In general, for the organic compounds considered on the first PSL, drinking water contributed little to total exposure.

To generate additional data relevant to the assessments for PSL 2 substances, a multimedia survey of 50 homes in Toronto has also been initiated. In this survey, the pilot phase of which has now been completed, samples of personal air, indoor air, outdoor air, drinking water and diet are being collected.

3.4 ASSESSMENT OF EFFECTS

The approach to assessment of the health risks associated with exposure to Priority Substances in the general environment, depends on the nature of the critical effect. Critical effects are defined as the biologically significant adverse effects expected to occur at the lowest dose or concentration. For many types of toxic effects (i.e., organ-specific, neurological/behavioral, immunological, epigenetic carcinogenesis, reproductive or developmental effects), it is generally assumed that there is a dose or concentration below which adverse effects will not occur (i.e., a threshold). For other types of toxic effects, it is assumed that there is some probability of harm at any level of exposure (i.e., that no threshold exists). At the present time, the latter assumption is generally considered to be appropriate only for mutagenesis and genotoxic carcinogenesis.

Priority Substances are classified into six categories based on the weight of evidence for carcinogenicity, including consideration of genotoxicity and mechanism of action.

Group I: carcinogenic to humans.
Group II: probably carcinogenic to humans.
Group III: possibly carcinogenic to humans.
Group IV: unlikely to be carcinogenic to humans.
Group V: probably not carcinogenic to humans.
Group VI: unclassifiable with respect to carcinogenicity to humans.

The criteria by which compounds are classified in the various categories range from convincing evidence in well conducted epidemiological studies in group I to inadequate or no data on carcinogenicity in group VI. This scheme is based in part on that developed by the International Agency for Research on Cancer (IARC). There are, however, several important differences. For example, the weight of incomplete epidemiological data is considered greater in the IARC scheme. Also, there is an additional category in the scheme for the classification of Priority Substances, namely 'unlikely to be carcinogenic to humans', to accommodate the increasing database on the mechanisms of carcinogenicity, particularly with respect to tumors induced in animal species which may be less relevant to humans. This scheme will likely be revised on the basis of experience gained in the assessment of PSL 1 and early PSL 2 substances.

Substances may also be classified into six corresponding categories on the basis of the weight of evidence for their potential to cause heritable mutations in humans, including consideration of the probability of the substance reaching the germ cells. At present, no comparable scheme for mutagenicity has been developed by other organizations.

3.5   DEVELOPMENT OF EXPOSURE POTENCY INDICES

For those substances for which the critical effect is considered to have no threshold (i.e., currently restricted to heritable mutagens and genotoxic carcinogens), data on the exposure–response or dose–response relationship in animals or humans exposed to high concentrations are often extrapolated using mathematical models to estimate risks at the much lower levels to which the general population may be exposed. There are numerous uncertainties in this approach, which generally involves extrapolation of results from animal experiments over several orders of magnitude, often in the absence of data on mechanisms of tumor induction or differences in toxicokinetics and toxicodynamics between the animal species and humans.

Quantification of risks for Priority Substances in terms of predicted incidence or numbers of excess deaths per unit of the population was considered inappropriate, primarily as this approach implies a degree of precision that is unwarranted in view of the numerous assumptions on which the low-dose risk estimates are based. Therefore, a risk measure, the Exposure/Potency Index (EPI), was developed for compounds classified in groups I or II (and occasionally group III) of the classification schemes for carcinogenicity or heritable mutations, primarily to provide guidance in establishing priorities for further action in the risk management stage. The EPI is a comparison of the estimated daily exposure of the general population (or certain high-exposure subgroups) to quantitative estimates of the carcinogenic or mutagenic potency of substances. Potency is expressed as the concentration or dose of a Priority Substance which induces a 5% increase in the incidence of, or deaths due to, tumors or heritable mutations (e.g., the tumorigenic dose or concentration 05, the TD_{05} or TC_{05}). It may be based on results of epidemiological studies, generally in occupationally exposed populations, or bioassays in experimental animals. In the latter case, the estimates of potency are restricted to effects considered relevant to humans. Estimates are based on tumors for which there has been a statistically significant increase in incidence and a dose–response relationship, using appropriate mathematical models (e.g., multistage).

Any model which fits the empirical data well is likely to provide a reasonable estimate of the potency. Choice of the model is not as critical as that for low-

Table 3.3 Exposure Potency Indices (EPI): Priority Substances List 1

Group I and II carcinogens	Risk (EPI)
Arsenic and its compounds	M, H
Benzene	H
Benzidine	L
Bis(chloromethyl) ether and chloromethyl methyl ether	L
Cadmium—inorganic compounds (inhalation)	H
Short chain chlorinated paraffins	insufficient data
Chromium VI (inhalation)	M–H
3,3'-Dichlorobenzidine	L
1,2-Dichloroethane	L–M
Dichloromethane (PBPK modified)	L–M
Hexachlorobenzene	M–H
Refractory ceramic fiber	L
Oxidic, sulfidic, and soluble nickel	M–H
Five polycyclic aromatic hydrocarbons	M, H
Trichloroethylene	L–M

L, low; M, moderate; H, high; PBPK, physiologically based pharmacokinetic (modeling).

dose extrapolation as estimates are based on doses within or close to the experimental range. The value of 5% is arbitrary; selection of another value close to the experimental range would not affect the relative magnitudes of the EPIs for each of a range of compounds. Wherever possible and if considered appropriate, information on toxicokinetics, metabolism, and mechanisms of carcinogenicity is incorporated into the quantitative estimates of carcinogenic potency, particularly those derived from studies in animals.

Compounds for which EPIs have been developed are assigned to one of three categories, high, moderate, or low priority for analysis of options to reduce exposure. For example, when estimated exposure is only a very small proportion of the concentration or dose which induces a 5% increase in tumors, the priority for analysis of options to reduce exposure is considered low. EPIs for the 15 compounds or groups of compounds on PSL 1 in carcinogenicity groups I or II were rather evenly divided among categories of high, moderate and low risk (Table 3.3).

3.6 DEVELOPMENT OF TOLERABLE INTAKES OR CONCENTRATIONS

For substances in groups III–VI of the classification schemes, for which the critical effect is considered to have a threshold, a dose or concentration that does not produce any (adverse) effect; i.e., 'no-observed-adverse-effect-level' (NOAEL) is identified where possible. Alternatively, a 'benchmark dose' or model-derived estimate of a particular incidence level (e.g., 5%), for the critical threshold effect can be estimated when appropriate data are available. If a value for the NOAEL or benchmark dose for the critical effect cannot be ascertained, tolerable daily intakes or concentrations are based on a lowest-observed-adverse-effect-level (LOAEL). The nature and severity of the critical effect and, to some extent, the steepness of the dose–response curve, are taken into account in the establishment of the NOAEL or LOAEL.

An uncertainty factor is applied to the NOAEL or LOAEL to derive a tolerable daily intake (TDI) or tolerable daily concentration (TDC), i.e., the daily intake or continuous concentration in air, respectively, to which it is believed that a person can be exposed over a lifetime without deleterious effect. Ideally, the NOAEL is derived from a chronic exposure study involving the most sensitive or relevant species, where possible, and utilizing data on species differences in toxicokinetic parameters or mechanism of action, or on investigations in the most sensitive subpopulation (e.g., the embryo or fetus in developmental studies). Wherever possible, the route of administration in critical studies in experimental animals should be similar to that by which humans are principally exposed. Generally, TDIs or TDCs are not developed on the basis of data from acute or short-term studies, unless observed effects in longer-term studies are expected to be similar. However, TDIs are occasionally based on data from subchronic studies in the absence of available information in adequately designed and conducted chronic toxicity studies; an additional factor of uncertainty is included in this case.

The uncertainty factor is derived on a case-by-case basis, depending principally on the quality of the database. Generally, factors of 1–10 are applied to account for both inter- and intraspecies variation. Where there is sufficient information, components of the factors for interspecies and intraspecies variation which address toxicokinetic (factors which influence exposure at the target site) and toxicodynamic (variations in sensitivity to the compound at the target site) differences are replaced with data-derived values. This is an approach which has been endorsed internationally by a task group of experts from the United States, Canada, the United Kingdom, several European countries and Japan convened to recommend to the International Programme on Chemical Safety (IPCS) methodology for developing guidance values in its Environmental Health Criteria Documents (IPCS 1994).

There were few compounds on the first PSL for which data on inter- and intraspecies variations in kinetics and dynamics were considered sufficient for replacement of default values in the calculation of TDIs/TDCs or for scaling of carcinogenic potencies. However, there was sufficient information to address interspecies variation in dynamics for peroxisome proliferation induced by diethylhexylphthalate (DEHP). In addition, data were sufficient to incorporate interspecies variations in kinetics for dichloromethane (and marginal for DEHP, styrene, and tetrachloroethylene). However, with these exceptions, the data were not sufficient to allow replacement of default values for intraspecies (interindividual) variations in either dynamics or kinetics with data-derived factors. This approach was helpful, however, in focusing attention on gaps in the available information which, if filled, would permit development of more appropriate TDIs or TDCs or estimates of carcinogenic potency.

An additional factor of 1–100 accounts for inadequacies of the database that may include lack of adequate data on developmental, chronic, or reproductive toxicity, use of a LOAEL versus a NOAEL, and deficiencies of the critical study. Potential interactions with other chemical substances commonly present in the general environment or dietary requirements of essential substances are also taken into consideration in derivation of the TDI or TDC. In rare cases, an additional modifying factor of 1–10 may be incorporated for effects such as teratogenicity or where there is some evidence of carcinogenicity which could be related to the critical effect. Numerical values of the uncertainty factor normally range from 1 to 10 000. Databases that require uncertainty factors greater than 10 000 are not used as their limitations preclude development of a reliable TDI or TDC.

The value of the TDI or TDC is compared with the estimated total daily intake of a chemical substance by the various age groups of the population of Canada (and in some cases, certain high exposure subgroups) or to values based on concentrations in relevant environmental media. If the estimated exposure approaches or exceeds the TDI or TDC, comparison of exposure values with the NOAEL or LOAEL on which the TDI or TDC is based is made to provide guidance in establishing priorities for further action, i.e., consideration of options to reduce exposure of the general population.

3.7 RECENT DEVELOPMENTS: DOSE–RESPONSE ASSESSMENT

Health Canada is currently developing methodology for application of the TD_{05} or TC_{05} as a basis for guidance values for the quality of various environmental media. In the interim, the relationship between TDs/TCs and low-dose risk estimates can provide the basis for practical application of $TD_{05}s/TC_{05}s$. For example, values based on division of the TD_{05}/TC_{05} by a margin of (5000–50 000) afford similar protection to those generally considered by various agencies to be 'essentially negligible' (i.e., $10^{-5}–10^{-6}$). (As $TD_{05}s$ were computed directly from the curve within or close to the experimental region, division by an additional factor of 2 would equate approximately to the lower 95% confidence level.)

For substances on the PSL 2, there will be increasing use of the benchmark dose as the measure of dose–response for non-neoplastic effects. As the TD_{05} is essentially a benchmark dose for carcinogenicity, the approaches for carcinogenic and non-neoplastic critical effects of Priority Substances are converging.

For several years, the validity of existing default values for uncertainty factors for Priority Substances has been investigated, including values applied when data are based on a subchronic rather than a chronic study or LOAELs versus NOAELs, or when core data sets are limited. One of the relevant projects involves consideration of effect levels in subchronic and chronic studies conducted under similar conditions for 30 substances to provide information relevant to extrapolations from subchronic to chronic and LOAELs to NOAELs.

The importance of examination of existing data to characterize distributions of components of uncertainty for assessment of noncancer effects is also the premise for a larger project. Relevant information will be compiled for review by experts with the objective of reaching consensus on recommendations for appropriate distributions to apply to risk assessments of noncancer effects. Where it is considered that currently available information is insufficient, strategies for acquiring the necessary data will be developed. This project will involve a staged approach to ensure that all relevant information is adequately considered, starting with coordinated compilation of relevant data and continuing with several workshops of experts.

3.8 RISK CHARACTERIZATION

Expression of the results of risk assessments for substances on the second PSL will be tailored to address the needs of the subsequent, but separate, risk management (i.e., strategic options) stage. It is anticipated that this will be accomplished in part by better characterization of exposure involving application of a wider range of approaches, more exposure scenarios and greater attempts to relate concentrations in the environment to sources.

The degree of confidence in the database on which the risk estimates are based, will also be better characterized. It is likely that this will take the form of a qualitative or if possible, quantitative measure, relative to other substances assessed in the program and will include clear distinction of, for example, confidence in assessments based on epidemiological studies in exposed populations versus toxicological data in animal species.

3.9 DOCUMENTATION

Owing primarily to the need to produce a large number of assessments in a limited time frame, products developed during PSL 1 were concise assessment reports and more detailed supporting documentation. Each assessment report, which is published in both official languages of Canada, includes an introduction, in which the scope of the review and the nature of peer review is clearly delineated, a detailed referenced summary of the critical data and a detailed referenced risk assessment including evaluation of the weight of evidence for critical effects, quantitation of risk and full description of assumptions and uncertainties. The supporting documentation, which contains a more detailed account of all relevant data, is made available on request. This approach was adopted based on experience acquired in a previous program which indicated that the audience for detailed supporting documentation was relatively small.

A similar approach has been adopted by the IPCS, in preparing Concise International Chemical Assessment Documents (CICADs). This program was initiated predominantly in response to increasing pressures internationally to assess larger numbers of substances in shorter time frames, specifically in relation to recommendations of the United Nations Conference on Environment and Development. Based on national supporting documentation, detailed summaries of critical data and draft assessments (i.e., the CICADs) will be prepared for international peer review and finalization. As this IPCS project involves many of the countries most active in the production of assessments, it is hoped that it will also fulfil a longer-term objective to improve the quality and consistency of supporting documentation internationally as a basis for risk assessment. Indeed, in view of the rather limited global resources to address chemical risks, it is clearly desirable to increase the range of chemicals being addressed internationally in such programs through improved coordination and trust.

3.10 CONCLUSIONS

The approaches described herein were developed to ensure consistency and defensibility in the assessment of risk to human health for Priority Substances under CEPA. These approaches continue to evolve to incorporate recent developments in methodology for health risk assessment and to better address the needs of risk management. These include: (1) a hierarchical multimedia approach to exposure estimation with the extent of characterization based on the margin between preliminary estimates of concentrations in the environment and effect levels; (2) the development of exposure potency indices for carcinogens in lieu of low-dose risk estimates; (3) increasing use of benchmark doses; and (4) incorporation of toxicokinetic and toxicodynamic data, where available, to modify traditionally adopted uncertainty factors for development of tolerable intakes, or concentrations, for non-neoplastic effects. These approaches have been shared with international programs to contribute to and benefit from international harmonization exercises designed to utilize global resources more efficiently in chemical risk assessment.

REFERENCES

Government of Canada (1996). Report of the Ministers' Expert Advisory Panel on the Second Priority Substances List under the Canadian Environmental Protection Act. Ottawa.

Environmental Health Directorate (1981). Tap Water Consumption in Canada. Report 82-EHD-80. Health and Welfare Canada, Ottawa.

Health Canada (1994). *Canadian Environmental Protection Act—Human Health Risk Assessment for Priority Substances.* Ottawa.

Health Canada (1996). *Health-Based Tolerable Daily Intakes/Concentrations and Tumorigenic Doses/Concentrations for Priority Substances.* Ottawa.

International Programme on Chemical Safety (1994). *Assessing Human Health Risks of Chemicals: Derivation of Guidance Values for Health-based Exposure Limits.* Environmental Health Criteria 170. World Health Organization, Geneva.

Nutrition Canada (1977). Food Consumption Patterns Report. Health and Welfare Canada, Ottawa.

4

Communicating Differences in Risk Approaches

CHRISTINE F. CHAISSON
Technical Assessment Systems, Inc., Washington DC, USA

4.1 INTRODUCTION

There are three masters that risk assessors serve: government, business, and the public. Risk assessors in the United States deal not only with state and federal agencies such as the Environmental Protection Agency (EPA) and the Food and Drug Administration (FDA), but also government regulators in the United Kingdom and Ireland, all through various parts of South America, Central America, Mexico, Japan, Australia, Malaysia, Philippines; indeed, around the world. More and more, these governments make demands based on their own risk assessment policies, which do not necessarily match those of the EPA and FDA.

The second sector, business, consists of organizations that require risk assessments in their daily activities. There is an obvious need for risk assessment by the producing industry, e.g., Dow Chemical and DuPont, which must satisfy concerns about the safety of packaging, waste disposal, new pesticides, new drugs, etc. There is a less obvious need for risk assessment on the part of bankers who are a major source of capital to industry and are the ones who extend lines of credit that support international trade. If internationally traded goods fail to meet some regulatory demand on the part of the recipient country, the bankers' investments are put at risk, and this provides significant incentives to conduct risk assessments.

The risk assessments businesses perform are the same kind that are submitted to the EPA, but they are used internally to support investment and other strategic decisions. These internal decisions have at least as much impact as those made by government regulators in determining the kinds of commodities available in the marketplace. Thus, business represents a silent, formidable, risk manager and enforcer.

Environmental Risk Harmonization: Federal and State Approaches to Environmental Hazards in the USA.
Edited by M. A. Kamrin. Published in 1997 by John Wiley & Sons Ltd, Chichester. ISBN 0 471 97265 7

Probably the most formidable force in the risk assessment process is the public. While the public is often thought of as the consumer or the consumer advocate, citizens have a very strong influence through the courts as a result of the awards of large settlements by judges and juries. As the United States is a very litigious society, potential liabilities associated with products have significant impacts on business. Detailed risk assessments are most frequently performed by companies in support of marketing decisions about new high-value products. When evaluating products that might be brought to market, those posing the least risk are most likely to be chosen. At a minimum, products presenting evidence of high risk will be viewed with deep skepticism by the business manager.

Service professionals represent another part of the public. These include health insurers, teachers, and the medical professionals. Citizens are more likely to rely on the local nurse, local pediatrician, or other local authority than the EPA for advice about health risks. Another very powerful professional group consists of politicians; while risk assessment has a science component, it is also heavily influenced by policy, tradition, and law.

4.2   RISK COMMUNICATION PRINCIPLES

An important question is how to communicate risk assessment and risk management between and among members of these sectors. This has been discussed at numerous meetings and it is obvious that all those who are performing risk assessments and risk communication, and making risk management decisions, have good intentions. They believe they are protecting public health and engaging constructively in an ongoing dialogue.

However, many risk communication difficulties remain and it is not clear if it is possible to communicate what risk assessors and risk managers have been doing for the last 30 years. While this is still an open question, there are some basic communication principles that have been uncovered. First, it is important to know the audience because communication strategies depend on whether the audience is the public, a science group, an industry or business group, or politicians. It also depends on whether audience members already have an opinion, or have heard conflicting information about the topic of discussion.

The second point is when a risk assessment opinion is proffered, whether it's qualitative or quantitative, the presenter needs to consider the audience's view as to his or her right to have an opinion. Has the presenter earned this right through trust? Or does the speaker have the right because of his or her power, e.g., the economic power of the banker or the enforcement power of the regulator. Although the presenter would undoubtedly prefer to be listened to because he or she was trusted, they should ask themselves whether they have the right to an opinion based on their power or position, or because the public genuinely needs to know their opinion on this topic. They should then act with this in mind.

Another question the communicator should ask is whether the message meets the expectations of the audience. The expectations of a science group are likely to be very different from those of the public. For example, a scientific audience may be more interested in the quality of the data and the analyses. However, even scientists may have biases about data depending on the source of the information, including the country in which the studies were performed.

If the communication involves recommended actions, then the consequences of these actions must be justified by the communicator, including considerations of social responsibility, e.g., equity. For example, a risk management decision to require industry to make safe fat replacers so that people can eat as many potato chips as they want is often preferable to the public to one suggesting they use a little more willpower. The public prefers regulations that put pressure on industry rather than management strategies that require personal responsibility. Indeed, the level of assurance that needs to be provided will be a lot greater if individual action is requested; the public is likely to be more accepting of uncertainties if it is 'big industry USA' that is asked to act rather than themselves. In sum, the consequences of the risk assessment message will have a significant impact on how it should be presented.

Two other risk communication factors are the content of the message and the context of the issues addressed. For example, it is important when communicating about natural ingredients and/or products to recognize that the public thinks natural ingredients are perfectly safe, while they usually hold the opposite view of genetically engineered products; although scientists would not necessarily agree. As a result, it is not surprising that bankers and insurance companies tend to be uncomfortable with products that are associated with DNA.

Communications about pharmaceuticals illustrate another issue; in this case, the public starts with the perception of benefit. As a result, people are willing to accept more risk, and a great deal of uncertainty. For example, if parents are told they cannot have a drug that might save their or their child's life until there is absolute certainty that it is safe, they will probably choose to live with the uncertainty and accept the potential risk. In contrast, however, they would be unlikely to accept that same level of risk or uncertainty with respect to a contaminant in the liner of their milk carton.

Further, the risk assessment message needs to include not only what was found but also how carefully the risk was studied. Did the investigator only look for obvious health impacts, or were measurements made using the most sensitive equipment? Or was the risk assessment a theoretical one based totally on modeling? For example, the European Union is using modeling to screen simultaneously 7000–20 000 different food additive products as part of new directives.

At the other extreme are intensive toxicology and exposure studies, such as those on the ethylenebisdithiocarbamates (EBDCs) that cost US$20 million. While the manufacturer may be required to pay for some studies of this type, who is going to pay for that level of testing for substances that are not regulated, e.g., aflatoxins? In some cases, an exporting country will have to supply such data if it wants to market its product in a receiving country. In most cases, however, the public will have to make decisions without the benefit of such data.

4.3   RISK COMMUNICATION AND RISK ASSESSMENT

There are a number of risk assessment issues that affect risk communication. In many cases, these issues reflect a lack of agreement about the conduct and interpretation of studies upon which risk assessments are based. These conflicts have negative impacts on risk communication as they often confuse the public and also make it difficult for citizens to place risk assessment results in the proper perspective.

There are a number of decisions that have be made regarding the appropriate performance of toxicity tests used in risk assessment. Examples include: How many animals to test, at what doses and for how long? What kinds of stains to use for pathology studies? What tumors should be counted? and, What enzymes should be measured? In addition, there may be questions as to whether behavioral effects tests should be performed and, if so, how should they be scored. There are dozens of these details, each of which is extremely important because together they determine the data that form the bases for the qualitative and quantitative assessments. Different countries require different kinds of protocols, and some of these protocols are mutually exclusive. For example, a rat cannot be sacrificed on day 13 as required by one country and then sacrificed again on day 14 as required by another nation.

Methodological issues such as these need to be resolved and there have been some steps taken in this direction, e.g., the acknowledgment by some countries that there were incompatible differences among them. In addition, some of these issues are moving toward resolution, although significant differences remain, e.g., the EPA in its latest cancer risk guidelines made significant steps to address outstanding issues with regard to the use of maximum tolerated doses. Thus, while harmonization on many of these issues is possible, it will probably take some time until all are resolved.

Assuming that the methodologies are harmonized, there is still the problem of consistent interpretation of the data generated. For example, should all the malignant tumors in test animals to be counted and then added to all the benign tumors, as well as all the hyperplasias? There are very strong opinions on these types of issues, and they are critical to whether a chemical is designated a carcinogen, a reproductive toxicant, etc., and whether it is ranked high risk or low risk.

Another issue affecting data interpretation is reproducibility of results. This can arise because the animals are different when the study is repeated, the power of the test is different in each instance or there are differences in expertise among the researchers. Suppose two identical tests were performed, one in Michigan and one in Germany, and one was positive and one was negative. In addition, the German study was better, but the American study was performed by a well-known professor. The fact that these factors influence data interpretations illustrates the importance of nonobjective criteria in risk assessment.

A third issue is the appropriate use of models in risk assessment. For example, when models are used to generate distributions for exposure parameters, e.g., exposure duration, what do the resulting population percentiles mean? Do these refer to the percentile of the entire population, the percentile of children, the percentile of Hispanics, etc.? Further, populations are not homogeneously distributed, so that 90th percentile of the population of the United States may easily equate to the 50th percentile of children.

A final interpretation issue is extrapolation. A good mathematics consultant can find an extrapolation model that will fit any purpose. This means that it is important to go beyond the result and to determine whether the extrapolation approach is scientifically valid. Unfortunately, modeling data have an advantage over experimental data because they generate smoother curves and thus look more believable. People have to be reminded that real data are messy and typically do not fit any theory precisely.

4.4   RISK COMMUNICATION AND RISK MANAGEMENT

In addition to methodological differences in risk assessment, there are philosophical differences in risk management. An illustration of this occurred about 7 years ago when I went to Germany to explain Proposition 65 and the Delaney Clause to a group of business people. At the end of the presentation, one of the Germans rose up and said, 'Dr Chaisson, how could you have permitted that to happen?' The audience could not understand how one can accept risk management approaches that are based on absolute risk standards. Explaining zero risk to a scientist or a lawyer, or especially to a banker, is very difficult. The current alternative to zero risk is to use 'bright lines' (or acceptable exposure limits) as the bases for risk management decisions.

Another philosophical question is what adverse effects are most important? The US public is very cancer-phobic; therefore, a great deal of legislation and a tremendous amount of resources are devoted to cancer studies. In most of Europe, however, reproductive risk is considered most serious and this effect receives as much attention as cancer does in the United States. In the Netherlands, there is also much more concern about worker exposure and worker risk issues.

The question of what segment of the population to protect in adopting various risk management strategies is also a philosophical one. Currently, the focus of risk regulation is to protect not only the average person, but also women and children. However, the same toxicological and exposure data that indicate that women and children are different from the average male also apply to the elderly, yet there is little social pressure to protect older individuals. This is reflected in both risk assessment and risk management decisions.

Another risk management issue is how to manage synthetic versus natural risks. Does it make sense to require soda to carry a label stating that it contains a possible carcinogen (or reproductive toxicant) when this requirement is not applied to coffee? The difference results from a risk management decision to consider the toxicants in coffee as natural constituents of the product while the same natural ingredients in the soda are considered food additives. What are the implications for a pregnant woman who has a choice between a can of soda with warnings about potential adverse effects and a cup of coffee without any warnings? Does this information communicate what she needs to know to protect her and her offspring's health?

A further philosophical issue is whether to base risk management decisions on risk versus benefit considerations. If this approach is used, it is important to determine who bears the risk and who gains the benefit. One possible impact of risk versus benefit considerations is risk shifting. An example is where a risk management decision reduces the risk for the general public but increases it for production or transportation workers. The latter individuals may be less vocal because of their circumstances of employment, and therefore this type of risk shifting may be considered acceptable.

A last risk management issue is whether to use risk versus risk comparisons in environmental decision making. This type of comparison can be difficult to make and communicate as it might involve comparing carcinogenic risks to adults with developmental effects in children and/or with ecological effects in birds. It is unlikely that risks that are so dissimilar can be compared objectively, and thus it is also very unlikely that everybody in society, including all three masters mentioned earlier, will be satisfied by the same decision.

4.5 CONCLUSIONS

Although a large number of difficulties have been discussed, it is encouraging that dialogue on these issues not only continues, but seems to be invigorated. The fact that—(1) conversations are currently being held in Europe addressing very fundamental issues, e.g., how to define a carcinogen; (2) US regulators are also having this conversation; and (3) that teachers and physicians and others want to be part of the conversation—is very positive, indeed. This will exert pressure on the organizations addressing harmonization issues, such as the Organization for Economic Cooperation and Development and the World

Health Organization, to move vigorously to reduce inconsistencies in risk assessment and risk management. However, for progress to continue, governments must support these efforts and resist the temptation of using risk assessments as trade barriers. It will make risk communication all the more difficult if the risk assessment tool is turned into a weapon that governments use against each other.

5

Risk Harmonization: A View from the Ground

DAVID A. BENNETT

*Office of Emergency and Remedial Response, US Environmental Protection
Agency, Washington DC, USA*

Risk harmonization means quite different things to different people. To begin, consider the term 'consistency' in risk assessment and risk management to be synonymous with harmonization; this definition will be refined as particular environmental issues are addressed. This initial definition reflects the most important goals in the Superfund Program: to build good risk assessment, to build consistent approaches to risk assessment, and to implement consistent risk management methods. The perspective of the author will be that of a laborer in the fields or a soldier in the trenches: a view from the ground.

5.1 INTRODUCTION

The beginning of a comprehensive effort to address existing hazardous waste sites was the passage by the US Congress of the Comprehensive Environmental Response, Compensation, and Liability Act of 1980 (CERCLA), the Superfund. This act was amended extensively in 1986 with the passage of the Superfund Amendments and Reauthorization Act (SARA). The US Environmental Protection Agency (EPA) was given the responsibility of carrying out this legislation.

This legislation includes two threshold requirements, two goals that the nation has to meet. One is to protect human health and the environment, and the second is to meet 'applicable or relevant and appropriate requirements', commonly known as ARARs. ARARs are requirements or standards found in other federal or state environmental statutes and the regulations promulgated to carry them out. An example of an ARAR is a maximum contaminant level (MCL) under the Safe Drinking Water Act. In the Superfund program, this MCL

This chapter is a US Government work and, as such, is in the public domain in the United States of America.

would be the cleanup goal for contaminated water in an aquifer designated to be used for drinking purposes.

In contrast to this simple example, there may be a number of different ARARs promulgated by various federal and state agencies for the same chemical in the same medium, e.g., soil. One state's standard may be risk based, another's, not. In light of these differences, it can be seen that CERCLA implicitly introduced conflict and inconsistency to the remediation process by mandating the use of ARARs. In response to this problem, the EPA has adopted a number of strategies over the years to minimize these inconsistencies and work down the path of harmonization.

One harmonization strategy has been the adoption of specific risk assessment and risk management approaches. For example, one important decision made in the Superfund Program was to use a quantitative site-specific approach to risk assessment, the so-called baseline risk assessment, to decide whether action is needed. This risk assessment approach is specifically designed to evaluate long-term risks. At least one of these risk assessments should be performed at every Superfund site on the National Priorities List. There are currently more than 1300 of these sites, and in the coming years the total might reach about 2000. While the approach is the same for all sites, some of these CERCLA sites are much more complex than the others, e.g., abandoned chemical facilities or smelters. With such great variability in site characteristics it is often difficult to achieve consistency in risk assessment from site to site and thus to make consistent remediation decisions based on a comparison of the risk assessment results to the ARARs that pertain to the site under consideration.

To address the problem of site variability the EPA has undertaken a number of harmonization efforts, i.e., attempts to improve consistency in assessing and managing the risk at different sites. An important first step was taken in 1989 with the publication of *Risk Assessment Guidance for Superfund* (RAGS) (US EPA 1989). This document has been the basis of all subsequent risk assessment efforts within the EPA and has also been used quite extensively by state programs in developing their approaches to assessing and managing risk at contaminated sites. The bases for the guidance in RAGS are the National Academy of Sciences National Research Council report entitled *Risk Assessment in the Federal Government: Managing the Process* and EPA guidelines (NRC 1983).

The RAGS document addresses both exposure and toxicity assessment methods. One way it attempts to improve consistency in assessing exposure is by placing heavy emphasis on criteria for performing site-specific exposure assessments, including those for extensive preparatory work, e.g., developing site conceptual models, establishing valid sampling protocols, and adhering to quality assurance requirements. Further, the guidance encourages assessors to examine a range of possible exposure factors.

Once the data have been collected, performing exposure assessments

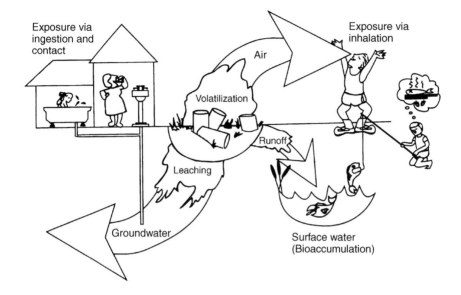

Figure 5.1   Assessing exposure: pathways.

requires a number of assumptions. Figure 5.1 illustrates a number of possible residential exposure scenarios and pathways at a hypothetical hazardous waste site. The guidance includes assumptions related to each pathway and also provides a set of intake equations that have been developed to calculate the exposure by each pathway. These equations are routinely used by the EPA, although they may be modified on a site-specific basis, and have also been adopted by a number of states for calculating exposures at local sites.

To illustrate how exposure assessment is done, one can consider a typical exposure scenario where chemicals on a site contaminate surface water. In this case, one route of exposure of concern is consumption of contaminated fish. To assess this, a survey of the people in the area might be undertaken to assess their fish consumption behaviors, e.g., what types of fish they eat and how much they eat. As the contaminants at the site are also likely to be found in other media such as soil, exposures from these other routes at the site would be evaluated as well using the appropriate equations and assumptions.

On the other hand, the toxicity assessment is not site specific. Instead, it is based on the EPA's Integrated Risk Information System (IRIS), which includes both hazard identification and dose–response components (US EPA 1993). Maximum acceptable daily intakes, known as Reference Doses (RfD), have been determined for many chemicals by the EPA and are published on IRIS. IRIS also provides slope factors and characterizations for carcinogens. When the chemical of concern has not been evaluated and an IRIS value published, the EPA Office of Research and Development provides assistance in calculating a toxicity value utilizing the same basic approach. This generic chemical toxicity information is combined with site-specific exposure information to perform a site-specific risk assessment.

After the risk assessment is completed, a decision must be made about possible remedial action. This decision is largely based on a comparison of the site-specific assessment value with the value that is considered an unacceptable risk by the EPA. In carrying out the Superfund Program the EPA decided that action is needed if the individual incremental lifetime cancer risk is greater than 1 in 10 000. The range of concern is from this value to 1 in 1 000 000, and the preference is to take actions that reduce the risk as close as possible to the lower end of the range. For noncarcinogenic endpoints, the standard for comparison is a hazard index of less than 1. This approach also takes into account possible exposure to multiple chemicals at a site.

In addition, the information gained from the risk characterization process is considered. This information helps explore uncertainties in the assessment and thus is more complex than a single risk value. Because of the complexity, the characterization may be difficult to communicate to staff people who make the cleanup decisions as well as to the public. This difficulty may result in risk managers and the public relying on discrete risk numbers and ignoring other information that has been gathered.

This other information reflects a variety of choices that are made in the process of exposure and toxicity assessments. For example, the risk characterization includes a description of who is being protected. The EPA's choice is generally to protect those at the high end of the distribution of exposures or, combined with toxicity, at the high end of risk values. These people often represent the 90th or 95th percentile of the population. This choice reflects a particular concern for sensitive subpopulations and highly exposed individuals. The characterization might provide information about average exposures as well.

In addition to comparing site-specific risk values with Superfund Program acceptable risks, these values must also be compared with the ARARs that have been developed by other governmental units. These ARARs may have been developed using different risk methodologies and assumptions and may include other considerations than those considered by the EPA. For example, some states have chosen background rather than a risk-based value as an acceptable environmental concentration in some media at a site. Other states have antidegradation laws for groundwater that prohibit adding anything to the groundwater that increases the contaminant concentration. These ARARs represent obvious conflicts between the EPA and state approaches, and they often result in a confused public.

As indicated earlier, the initial RAGS document of 1989 was designed to address this conflict and to create a standard approach. It was soon apparent that this was just a start and additional volumes of guidance with a similar goal were developed and published. However, the net result was not the hoped for standard or consistent assessments because there were many different approaches taken to implement the guidance.

The EPA response was to issue more guidelines, e.g., EPA (1991a). In addition, a memorandum describing standard default exposure factors was published (US EPA 1991b). This did result in more consistent implementation, but also at times led to another problem. This was rote implementation, where the guidance was implemented using all the default values, even where they may not have been appropriate. These problems in the EPA's attempts to harmonize risk assessments often resulted in different approaches being applied at different sites or in inappropriate approaches being used at some sites. This showed that consistency in guidance may not achieve consistency in implementation.

5.2 RECENT APPROACHES

The EPA has continued to issue guidance and tried to work toward consistency. One recent approach was to streamline consideration of soil contamination. Concentrations of contaminants in soil can range greatly from zero or very low up to very high values. Clearly, at the low end there is some concentration that is not of concern, and at the high end there is a level that requires immediate action. In the middle is where most of the difficult decisions must be made. Recognizing this, the EPA issued the *Soil Screening Guidance* utilizing an approach that, as a first step, identifies soil levels below which there is generally no concern (US EPA 1996).

These screening levels are not meant to be cleanup goals; rather they represent concentrations above which more study is needed at the site. This screening approach can be applied to a whole site or parts of a site and can be used to identify areas of little concern as well as those to which significant resources should be devoted.

The EPA chose to calculate soil screening levels on a chemical by chemical basis and for individual exposures. In determining these levels, a 1 in 1 000 000 risk standard was used as the risk criterion for carcinogenic chemicals as opposed to the 1 in 10 000 to 1 in 1 000 000 criterion used for cleanup decisions in the Superfund Program. The soil screening values were calculated based on direct ingestion of soil and groundwater by residential users. This approach was taken based on the assumption that if the levels are acceptable for residential use, then they are acceptable for all other uses. Inhalation of vapors and particulates is also considered in establishing soil screening levels.

The screening level approach also differs from previous guidance in that it involves a tiered approach. The first tier is to calculate an acceptable concentration of a particular contaminant in the soil using equations and default factors such as those shown in Figure 5.2. The result is a concentration in soil in parts per million that is protective of young children. The second tier is a

$$\text{Screening level} \atop \text{(mg/kg)} = \frac{\text{THQ x BW x AT x 365 day/year}}{\frac{1}{\text{RfD}_o} \times 10^{-6} \text{ kg/mg x EF x ED x IR}}$$

Parameter/Definition (units)	Default
THQ/target hazard quotient (unitless)	1
BW/body weight (kg)	15
AT/averaging time (year)	6[a]
RfD_o/oral reference dose (mg/kg-day)	chemical-specific
EF/exposure frequency (days/year)	350
ED/exposure duration (year)	6
IR/soil ingestion rate (mg/day)	200

[a] for noncarcinogens, averaging time = exposure duration

Figure 5.2 Screening level equation for ingestion of noncarcinogenic contaminants in residential soil.

simple site-specific approach that does not use all default values, but rather certain EPA recommended values for soil. These values incorporate characteristics of the soil that particularly affect either contaminant volatilization or the transport of contaminants through soil to groundwater. The third tier is a full-blown risk assessment. The first two tiers are solely human health based and address only the most common residential exposure pathways. It is important to recognize that there also may be risks to the environment, e.g., wildlife, or to people through less common activities such as raising beef. In this case, the rote approach may not be appropriate and a site-specific approach should be used.

5.3 DISCUSSION

What has been learned about harmonization as a result of attempts to implement CERCLA? One lesson that resulted largely from the Soil Screening Guidance process is the importance of communicating with a wide range of people in creating guidance. The Soil Screening Guidance process included input from risk assessors, state agency staff, industry representatives, bankers, insurance people, and environmental groups. Each brought a different perspective to the process. For example, one concern expressed by many industry representatives was that providing a table of generic screening levels might inevitably lead to their misuse as cleanup values. On the other hand, a common environmental group concern was that people will use the table in rote ways and ignore very sensitive site-specific pathways, e.g., people who grow vegetables in their household gardens, or ecological targets. To address these concerns, the guidance emphasizes the use of the simple site-specific approach rather than just the tabular values. In addition, the table was published as a separate appendix rather than as an integral part of the guidance. It is not clear how effective these steps will be in preventing misuse of the published levels.

A second lesson is that implementation has to be responsive to changing regulatory goals and approaches, such as those happening in the states right now. These changes often include increasing the importance of balancing cost-effectiveness against risk. For example, one state that required cleanup of contaminants in groundwater to background levels now is developing three different cleanup options, although one option would still require remediation to background. Another state used risk-based decision making for cleanups, but its risk assessments were very stringent compared with the standard EPA approach. This state kept its basic approach, but recently lowered its criterion for carcinogenic risk from 1 in 1 000 000 to 1 in 100 000, and added flexibility in the application of exposure factors. A third state seems to be moving in the other direction towards a more stringent 1 in 1 000 000 cleanup requirement. A fourth state is still wrestling with its groundwater antidegradation policy and how to deal with that requirement in its site remediation process. Thus, any successful guidance has to have the flexibility to respond to such state changes.

In addition to these actions at the state level, changes are also occurring at the national level, e.g., the adoption of an alternative approach called risk-based corrective action (RBCA) for leaking underground storage tanks. This approach was generated by the EPA Office of Underground Storage Tanks, working through the American Society for Testing and Materials (ASTM). As a result of this process, an ASTM standard has been issued (ASTM 1995). This standard provides a model risk management approach for investigating and remediating leaking underground storage tanks that is fundamentally based on risk assessment guidance for Superfund.

From the point of view of harmonization, another interesting aspect of RBCA is how this approach is being implemented. The EPA has entered a public/private partnership with industry that is training the states to use this approach. To accomplish this, training sessions need to discuss the various aspects of the risk assessment process, such as the RBCA framework, methods for site investigations, exposure pathways, toxicity endpoints, etc. However, RBCA cannot provide a standard risk management approach as each state has to make its own judgement as to acceptable risk. What the trainers are finding, on the basis of experience with almost 40 states, is that many states have never confronted some of these questions before. In addition, different agencies or departments in states often have different views about acceptable risk. This training encourages states to address risk across programs in a generic fashion, not just with respect to underground storage tanks. As a result, the decision the state makes will often be applied to activities that might impact the Superfund process or a state voluntary cleanup program, or actions taken under the Resource Conservation and Recovery Act (RCRA).

Also of interest is that state participants often ask the trainers what is being done in other states. As a result, the trainers are putting together a database about the decisions that are being made in the different states. It is not clear what this database will show or how influential it will be in the various states.

Another example of activities at the national level related to harmonization is the recent EPA-sponsored Monte Carlo Workshop. This workshop, bringing together statisticians and other interested parties, addressed possible EPA approaches to incorporating probabilistic analyses, e.g., Monte Carlo simulations, into EPA risk assessment and risk management decisions. There were many suggestions including one to establish what might be called default exposure distributions as an alternative to the current EPA default point estimate approach under Superfund. The current approach utilizes a reasonable maximum exposure scenario in which, for example, it is assumed that children eat two-tenths of a gram of dirt per day, and children under 6 years play in their own backyards (US EPA 1991b).

To change from such discrete assumptions to an exposure distribution leads, again, to a harmonization challenge. Who will decide on the appropriateness of these distributions as they are developed? On what criteria will the decisions be based? In addition, when it is decided that not enough is known to make the best decision, who will support improving the data with research dollars and skilled investigators?

REFERENCES

American Society for Materials and Testing (1995). *Standard Guide for Risk-based Corrective Action Applied at Petroleum-release Sites.* E1739-95. West Conshohocken, PA.
National Research Council (1983). *Risk Assessment in the Federal Government: Managing the Process.* National Academy Press, Washington, DC.
US Environmental Protection Agency (1989). *Risk Assessment Guidance for Superfund,* Vol. 1. *Human Health Evaluation Manual* (Part A). Interim Final. EPA/540/1-89/002. Office of Emergency and Remedial Response, Washington, DC.
US Environmental Protection Agency (1991a). *Risk Assessment Guidance for Superfund,* Vol. 1. Part B. Development of Risk-based Preliminary Remediation Goals EPA 540/R-92/003. Office of Emergency and Remedial Response, Washington, DC.
US Environmental Protection Agency (1991b). Human Health Evaluation Manual, Supplemental Guidance: Standard Default Exposure Factors. OSWER Directive 9285.6-03. Office of Solid Waste and Emergency Response, Washington, DC.

US Environmental Protection Agency (1993). Integrated risk information system: announcement of availability of background paper. *Federal Register* **58**: 11490–11495.

US Environmental Protection Agency (1996). *Soil Screening Guidance: User's Guide.* EPA/540/R-96/018. Office of Emergency and Remedial Response, Washington, DC.

6

Risk Assessment: What is the Question?

P. MICHAEL BOLGER

Center for Food Safety and Applied Nutrition, US Food and Drug Administration, Washington DC, USA

The disharmony in risk assessment and risk management approaches used by various federal and state agencies is many times more apparent than real. The approach used depends on the question that is being asked, which in turn, reflects the legislative authority under which the assessment is being performed. Statutory mandates reflect a number of factors other than risk, such as social issues (e.g., risk perception), economic factors, and others. Thus, achieving consensus about risk assessment is not going to solve the problem of disharmony in risk management because many times the sources of disharmony arise not from differences in risk assessment approaches, but from the underlying differences in science policy, legislative authority, and risk standards among different statutes.

Another source of disharmony or, alternatively, barrier to harmonization is that, even though risk or safety assessment for the protection of public health is performed similarly in different agencies, there are a number of different names given to similar products of these assessments. The acceptable daily intake (ADI) was the first term coined to describe a 'safe' level of exposure and this was followed by the tolerable daily intake (TDI) of the World Health Organization, the reference dose (RfD) of the Environmental Protection Agency, and the minimal risk level (MRL) of the Agency for Toxic Substances and Disease Registry. All of these are different names for a value that reflects the same basic safety assessment approach and which generally reflects utilization of the same scientific base.

This chapter is a US Government work and, as such, is in the public domain in the United States of America.

6.1 THE FOOD AND DRUG ADMINISTRATION

The focus of this presentation is the risk assessment paradigm used in the Center for Food Safety and Applied Nutrition, Food and Drug Administration (FDA) and shown in Figure 6.1. The initial step is the safety assessment phase and corresponds to the risk question that an ADI, RfD, MRL, or TDI is designed to answer, namely what is a safe level of exposure? In most circumstances that information is all that is needed to resolve a public health issue as these safety assessment values represent exposure levels above which individuals or populations are said to be at risk. If they are not exceeded, then mitigation is not necessary.

However if exceedences occur, these values do not provide any quantitative information about the risk associated with higher than acceptable levels of exposure. Estimating the magnitude of such risks is important for the risk manager who must decide on risk priorities and the amount of resources to devote to each risk scenario. In light of this, the FDA approach to food-borne contaminants is, first, to perform a safety assessment and then move to a quantitative evaluation of risk when it is determined that human exposure to a dietary contaminant is unsafe.

The Federal Food, Drug and Cosmetic Act (FFDCA) specifies the standards of safety or risk that apply to contaminants in foods. These standards differentiate between added and naturally occurring contaminants. Section 402(a)(1) of the Act uses the phrase 'may render injurious to health' when referring to an

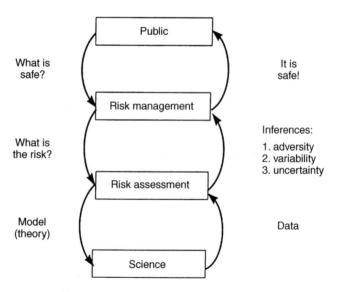

Figure 6.1   Risk assessment paradigm.

added substance. In this case 'added' refers to the hand of man. The Act uses another phrase, 'ordinarily render', to refer to naturally occurring toxic substances, like the mycotoxins vomitoxin or fumonisin, where the hand of man is not evident. The law thus designates qualitatively different criteria and acceptable levels of risk in the two cases. However, the Act does not specify what those levels of risk are.

While the distinction between anthropogenic and natural dietary contaminants is clear in the law, assigning risk values to each is difficult in practice because there are no established science policy guidelines for each type of contaminant. Thus, there are continual controversies about what risk standard applies to a particular contaminant. For example, some people interpret the 'ordinarily render' phrase as requiring a 'body bag' criterion for risk, i.e., no remedial action can be taken without documented and precise (e.g., unequivocal dose–response information) reports of human adverse reactions. Discussions about risk standards have become particularly relevant recently because of concerns about potential hazards of some dietary supplements.

The other critical section of the FFDCA is Section 406 which describes how tolerances are to be established for environmental contaminants and which uses the same 'may render injurious' risk standard that was discussed previously. However, it also specifies that the tolerance setting process must address the issue of avoidability and the limitations of current analytical capabilities. The rationale behind the latter consideration is that there is no point in setting a standard below the limit of what can be detected. In addition, the magnitude of the expected food loss and the prospect of near-term technological changes that would result in reductions in exposure must also be considered in tolerance setting.

6.2 SAFETY ASSESSMENT APPROACHES

The process of safety/risk assessment requires determination of the severity and scope of adverse effects of a hazard (toxicity assessment) and also the degree of exposure of an individual or population (exposure assessment). The current approach to toxicity assessment can be traced to a paper by Lehman and Fitzhugh in 1954 which introduced the use of 10-fold safety factors, which later also became known as 'uncertainty factors' (Lehman and Fitzhugh 1954). These factors are used to extrapolate data on the effects of specific chemicals from laboratory animals to humans; in particular, the no-observed-adverse-effect-level (NOAEL) is divided by the safety factors in determining the safe level. The resultant values are those discussed earlier, e.g., the ADI or RfD. It is important to note that the overall 100–1000-fold safety/uncertainty factor was described by Lehman and Fitzhugh as a target, not an absolute yardstick, as they concluded there were no scientific or mathematical means by which absolute values for these factors could be derived. However, in recent years these factors have been routinely used, not as targets but as accepted values which attempt to ensure an adequate margin of safety.

This safety factor approach does not attempt to describe uncertainty; instead it allows calculation of a safe value that includes an adequate margin of safety or, as it is sometimes known, an adequate margin of exposure. However, the safety/uncertainty factors used to calculate a safe exposure level are based in large part on what is not known and not on what is known of the toxicology of a contaminant. Indeed, the less that is known, the greater the total uncertainty factor applied.

One other important feature of the safety assessment approach is the way that it uses dose–response information. For example, two chemicals may have very different dose–response curves, but similar NOAELs. If the NOAELs are similar, then the resulting safe exposure levels (e.g., ADIs) will also be similar, and therefore, no distinction will be made between the two chemicals in regards to their potential risk. If dose–response was considered as well as the NOAEL in the quantitative derivation of the safe level, it is possible that a very different conclusion would be drawn regarding comparative risks of exposure to these two chemicals.

There is no single source, e.g., journal article, that describes the exposure assessment process and, indeed, such assessments are performed differently in state and federal agencies. As these agencies often have medium-specific mandates, e.g., drinking water, their exposure estimates reflect concerns about routes of exposure from these media and do not take into account the combined effect of all the possible sources and routes of exposure. This latter point is particularly important in dealing with ubiquitous environmental contaminants which may occur in a number of media as an assessment that addresses only one route of exposure may result in an erroneous evaluation of the significance of the risk from this contaminant.

Another critical issue is the use of the point estimate (or single value) approach in assessing exposures. In this approach, the overall estimate of exposure is calculated from point estimates of individual exposure factors, e.g., volume of drinking water ingested/day, body weight, etc. As a general rule these point estimates of exposure are conservative; they are based on the behavior of the 90th or greater percentile of individuals to accommodate concern for sensitive subpopulations or especially highly exposed consumers. With acute endpoints, particularly when dealing with possible life-threatening outcomes, even higher percentile levels of consumption/exposure are sometimes used to provide for a larger margin of exposure. In essence, the use of point estimates is a screening or prioritizing methodology. By utilizing conservative assumptions, this approach helps to distinguish problems from nonproblems and gives a rational basis for focusing resources on those exposures which may pose the greatest risk.

However, this safety assessment approach, utilizing uncertainty factors and conservative exposure assumptions, should be considered only the first step in an iterative process. As portrayed in Figure 6.1 it is useful for screening out trivial public health problems, but it does not provide an accurate description of risk and so does not provide a meaningful basis for gauging the level of effort needed to minimize or avoid exposures that may be of concern.

6.3 RISK ASSESSMENT

Once a possible problem has been identified through a safety assessment, the next step is to perform a quantitative risk assessment. An important part of such an assessment is the identification and estimation of major uncertainties that may be associated with the assessment. Let us first examine the exposure assessment component of risk assessment. There can be significant uncertainty when using point estimates of consumption along with contaminant residue data to estimate exposure quantitatively. For example, it is generally assumed that the population of concern consumes the same amount of a food containing the same level of a contaminant every day for a period of time. The reality is that most people do not eat the same types of food containing the same level of a chemical every day for prolonged periods of time. The types and amounts of food consumed vary and so resulting contaminant exposures will also vary. Differences in timing and duration of exposure can have a significant impact on the characterization of exposure and risk. For many dietary contaminants where exposure is intermittent the point estimate approach results in an overestimation of exposure and the resulting risk.

Another key uncertainty arises in the toxicity assessment, specifically in the interpretation of adverse outcome measurements in animal or human studies. For example, the implications of observing a particular adverse effect in an animal bioassay for human populations is not clear-cut. There is uncertainty related to the variability of response within an animal or human population, to differences in animal and human target organ sensitivity, and to differences in intra- and interspecies metabolism and toxicokinetics. In addition, uncertainties are introduced when models are used to extrapolate below the observed dose–response range as is done in carcinogen risk assessment.

Quantitative risk assessment should provide the risk manager with all of the relevant information, including uncertainties, in both exposure and toxicity assessments. It should also go beyond the assessment of one particular hazard and provide the decision maker a way to compare the risk of this hazard with other hazards, e.g., alternate foods. For example, if a population is advised to avoid a particular food because of concern for a chemical contaminant, the risk manager should be informed about other risks this population may face in consuming alternate food sources; especially, microbiological risks. A decision maker utilizing such a comparative analysis will have a better chance of instituting policies that will result in an increase in overall public health.

Comparisons of this kind, however, are difficult to make. For example, there are questions as to the best ways to assess microbiological risks. There is also the issue of how to measure doses of microbiological contaminants. Further, it may be helpful to consider nutritional benefits as well as chemical or microbiological risks but it is not clear how best to make this comparison. The difficulty in addressing these issues suggests that it is necessary to go beyond the current approach to risk assessment and management to achieve the best results. For example, as a matter of public policy, it is important to provide the risk manager and the consumer with more than a statement of the risk of eating fish containing a particular contaminant at a specific level and another qualitative statement about the general benefits of fish consumption.

6.4 THE LEAD EXAMPLE

Dietary lead exposure is a useful example of how the safety/risk assessment paradigm shown in Figure 6.1 is applied by FDA and how some of the issues raised may be addressed. Blood lead, the commonly used measure of dose or body burden, reflects a number of different inputs or sources and routes of exposure. If the public health concern is for lead in a particular food, the important question is how the magnitude of exposure from this one source of lead compares with that from other dietary lead sources and how it compares with the background body burdens of lead. If the focus of the safety/risk assessment is solely on the one dietary source or on food in general without considering quantitatively how it compares with the other sources, then the evaluation of risk is incomplete.

To provide this limited information to a risk manager or decision maker could very well result in decisions that yield insignificant improvements in public health but that are very costly to society and limit available resources for other public health endeavors that may have significant public health benefits.

Table 6.1 Total tolerable daily intakes for dietary lead (Pb)

Population group	Blood Pb level of concern (µg/dL)	Conversion factor[a]	Dietary effect level (µg/day)[b]	Total tolerable daily intake (µg/day)[c]
Children	10	0.16	60	6
Pregnant women	10	0.04	250	25
Adults	30	0.04	750	75

[a]Conversion factors empirically derived.
[b]Dietary effect level is equal to the blood level of concern divided by the conversion factor.
[c]Tolerable daily intake is equal to the dietary effect level divided by a safety/uncertainty factor of 10.

Worse yet, such an analysis could result in decisions that have significant negative, although unintended, consequences. A good example of this would be a decision to encourage individuals or populations to change their dietary habits because of concern for lead as a dietary contaminant; a decision that may increase their nutritional or dietary microbiological risks.

With lead the toxicity assessment is not based on cancer risk and thus on low dose extrapolation from an animal cancer bioassay; rather risk estimates are based on studies of humans who have been exposed to environmental lead. These data suggest that there is no observable threshold for lead toxicity and thus no margin of safety can be established. This means that in the case of lead it is necessary to go beyond a typical safety assessment in which safety/uncertainty factors are applied to a threshold dose, and to determine, quantitatively, the risk associated with different levels of exposure.

At present, management of risk from lead exposure is based not on exposure limits but rather on blood lead levels. Table 6.1 shows the blood lead levels of concern for several age groups as well as the calculated steady-state dietary lead intake levels that correspond to these blood concentrations (Carrington and Bolger 1992). These values are total tolerable daily intake levels for lead that could result from a single source or from all possible sources of lead. They cannot be defined as 'safe' levels in the classical sense in that they were not derived by the application of safety/uncertainty factors to a NOAEL, but rather from a human effect level.

A simple approach to safety assessment for a particular food and a specific sensitive subpopulation (e.g., pregnant women, children) is to compare a point estimate of lead exposure due to this food with the total tolerable intake level. This type of approach may be tenable if this food is the sole source of contaminant exposure. However, with lead this is almost always not the case and background lead exposure must be considered as well as other sources and pathways.

One approach that has been used (e.g., federal drinking water standards) for such a case is the relative source contribution (RSC) method in which sources are assigned a percentage of the total tolerable intake for a contaminant. A major drawback of this method is that the assignment of the percentage is generally arbitrary and not based on specific exposure information. This kind of approach often leads to tolerable intake levels for some sources that are not technologically or economically achievable. The inevitable conclusion of such an approach is that exposures like lead are unsafe. As lead exposure is inevitable, this is just a starting point and should be followed by an attempt to describe, quantitatively, the levels of risk associated with different expected exposure scenarios.

One method that can be used to derive such estimates is the Monte Carlo simulation. This is a probabilistic approach in which the input variables (e.g., food consumption data, lead levels in different foods and other media like water, dust, and soil) are described as distributions. A major advantage of this approach is that rather than selecting a single data point (e.g., mean) from a set of data, the whole data set is used. In this simulation technique, a large number of independent samples are randomly selected from the distributions of different variables (e.g., wine consumption, lead wine levels) and the corresponding outputs calculated to model how people may actually be exposed. Using inferential statistical methods, a series of output values are generated which provide the estimated distribution for the calculated parameter, e.g. wine consumption.

The probabilistic approach more accurately models the way people are actually exposed to a dietary contaminant such as lead. The point estimate approach assumes that each member of a population consumes the same amount of food containing the same level of a contaminant for a given period of time. In reality, individual day to day exposure is variable and this is reflected in the probabilistic approach that takes such variability into account.

An example of the use of this technique is a probabilistic assessment of dietary lead exposure, including lead in glazed ceramicware, drinking water, wine, and calcium supplements (Carrington *et al.* 1996) that was performed in FDA. This analysis provided a way to understand the risk from each dietary lead source superimposed on the background lead risk in a population. For example, if the concern is lead in wine consumed by women of child-bearing age, this approach addresses the question of how the lead in this one food contributes to the total body burden of lead in these women, a burden which reflects not only lead in the rest of the diet but also lead from nondietary sources.

Table 6.2 illustrates the many and different types of source consumption/

Table 6.2 Distribution of lead (Pb) levels in various sources

Pb source	Distribution type	Mean	Range	Confidence level
Air (µg/ml)	empirical	0.13	0–2	medium
Soil/dust (p.p.m.)	empirical	355	10–20 000	low
Tap water (p.p.b.)	lognormal	9.4	1–125	low
Diet (p.p.b.)	lognormal	9.1	2–100	low
Canned Food (p.p.b.)	lognormal	25	10–500	medium
Hollowware (p.p.b.)	lognormal	64	2–2000	very low
Flatware (p.p.b.)	lognormal	9	2–500	very low
Cigarette (µg/cig)	normal	0.15	0.05–0.25	low
Calcium supplement (µg Pb/g Ca)	empirical	9.2	0–50	high
Wine				
decanted (p.p.b.)	empirical	155	4–1980	high
undecanted (p.p.b.)	empirical	155	2–793	high

Table 6.3 Added risk for pregnant women of lead (Pb) from hollowware[a]

Target blood lead level	7 μg Pb/dL	10 μg Pb/dL
Background risk[b]	4.5×10^{-3}	2.1×10^{-3}
Added risk[c]		
5.0 mg/L	4.6×10^{-1}	3.8×10^{-1}
0.5 mg/L	1.8×10^{-2}	6.5×10^{-3}
0.1 mg/L	2.0×10^{-4}	2.0×10^{-4}

[a]Pb limits in ceramicware are based on tests which measure leaching into acetic acid for 24 hours.
[b]Risks based on background exposures (diet, air, water, soil, dust) and corresponding body burdens.
[c]Added risks based on simulations with fixed Pb levels to correspond several Pb limits.

intake data that were used in this analysis. The range and confidence level of the data sets clearly show how they vary in quality and quantity. This is addressed in the analysis by quantitatively describing the variability associated with each data set. In turn, the quantitative and relative impacts of each variable or data set on the uncertainty of each of the risk estimates is transparent and can be judged. The same is true for the range of levels of lead in the different sources.

The results of the Monte Carlo simulation of the added risk to pregnant women from lead leaching from ceramic hollowware is presented in Table 6.3. Depending on the target blood lead level of concern, whether it is 7 or 10 μg/ dL, an estimate of the background risk of lead can be derived. Then the added risk of exposure to lead from a particular source, in this case lead from ceramic hollowware, is estimated. As part of this analysis the impacts of three different plausible regulatory standards for lead migration from hollowware were considered. This provides an appreciation of how the three different standards change the risk of lead exposure from this one source.

The decision maker/risk manager can then judge the utility of successively lower standards in terms of risk reduction and determine the point at which further reductions become minimal. This information can then be balanced against the other factors (e.g., competing dietary risks, nutritional risks) that have to be considered in developing a rational and equitable public health strategy.

6.5 SUMMARY

In summary, in considering the risk of an environmental contaminant the first issue that must be addressed is a clear and unambiguous statement of the public health question the risk assessment will be used to answer. Is the question one of a safe level of exposure where the risk is negligible (essentially zero) or is it a question of numerical estimates of risk associated with different levels of exposure? It is important in any consideration of risk that those involved, the risk manager(s) and risk assessor(s), have a clear understanding of each others role and are communicating and not talking past each other. In the context of the risk assessment paradigm in Figure 6.1, they should be on the same level in the paradigm. If the public health question is one of quantitative description of risk, then the variability and uncertainty of the estimates will have to be dealt with quantitatively. Exposure for most environmental contaminants is multi-route, and varies in frequency and level over time. Uncertainty estimates influence every step of the risk assessment process, and a knowledge of these uncertainties provides the risk manager information needed to assess the reliability of the estimates of risk. In turn, this will give the risk manager the ability to compare estimates of risk and identify mitigation efforts that result in the greatest reductions of risk while utilizing limited public resources most efficiently.

REFERENCES

Carrington CD, Bolger PM (1992). Hazards of lead in food. *Regulatory Toxicology and Pharmacology* **16**: 265–272.

Carrington CD, Bolger PM, Scheuplein RJ (1996). Risk analysis of dietary lead exposure. *Food Additives and Contaminants* **13**: 61–76.

Lehman AJ, Fitzhugh OG (1954). Ten-fold safety factors. *Association of Food and Drug Officials of the United States, Quarterly Bulletin* **XVIII** (1): 33–35.

Role of Risk Assessment in Public Health Practice

**CHRISTOPHER T. DE ROSA, YEE-WAN STEVENS AND
BARRY L. JOHNSON**

*Agency for Toxic Substances and Disease Registry, US Department of Health and
Human Services, Atlanta, GA, USA*

7.1 INTRODUCTION

The Agency for Toxic Substances and Disease Registry (ATSDR) is one of eight
Public Health Service agencies within the US Department of Health and Human
Services. ATSDR was created by the Comprehensive Environmental Response,
Compensation, and Liability Act (CERCLA), also known as the Superfund. The
mission of ATSDR is to prevent exposure and adverse human health effects and
diminished quality of life associated with exposure to hazardous substances
from waste sites, unplanned releases, and other sources of pollution present in
the environment (ATSDR 1994a). To fulfil this mission, ATSDR conducts
programs in health assessment, epidemiology, health surveillance, applied
research, health education, and toxicologic database development. Risk assessment principles and practices have influenced the way ATSDR conducts these
programs. This chapter describes the influence of risk assessment on agency
programs, and ATSDR's departure from some traditional aspects of risk assessment.

In pursuing its legislatively mandated responsibilities, the agency addresses
public health concerns associated with multiple types of exposure to a wide
range of substances found at hazardous waste facilities. To accomplish these
responsibilities, ATSDR engages in the following program areas:

1. public health assessment of hazardous waste sites;
2. health consultations regarding specific hazardous waste sites and sub-
    stances;
3. health investigations;

This chapter is a US Government work and, as such, is in the public domain in the United States of America.

4. development of exposure and disease registries;
5. emergency response to releases of hazardous substances;
6. applied research to enhance its public health assessment capabilities;
7. identification, prioritization, and critical assessment of hazardous substances;
8. education and training of health care providers and communities potentially exposed to hazardous wastes.

To address the information needs associated with these programs, ATSDR
has developed a range of toxicologic and human health databases under the
mandates of CERCLA, as amended (Johnson 1995).
    Despite the existence of these databases, as well as a wide range of applied
research efforts to address specific data gaps, missing information continues to
confound efforts to characterize, fully and accurately, the public health implications of exposure to hazardous wastes. To meet this challenge, a wide range of
environmental programs have invoked the risk assessment paradigm as defined
by the National Academy of Sciences (NAS) (National Research Council 1983).
Risk assessment has been used to define the problem, to present information,
to evaluate available data, and finally, to formulate conclusions that integrate
what is known regarding exposure and potential health effects.
    As described by the NAS, risk assessment consists of four interrelated
components: hazard identification, dose–response assessment, exposure assessment, and risk characterization (Figure 7.1). Hazard identification addresses the
question of causality in a qualitative sense; that is, the degree to which evidence
suggests an agent elicits a given effect in an exposed population. Typically, this
first component of risk assessment includes: an evaluation of the validity of

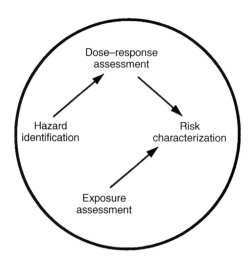

Figure 7.1 National Academy of Sciences/National Research Council risk
assessment paradigm (De Rosa *et al.* 1993).

toxicity data; a weight-of-evidence summary of the relationship between the substance and toxic effects; and estimates of the generalizability of data to exposed populations. Although the presumption is usually made that effects observed in a study population are generalizable to other populations, this concordance must be carefully examined in risk assessment.

The second component of the risk assessment, the dose–response assessment, provides an estimation of the relationship between the dose of a particular agent and the incidence of effects in a population. It is a method of establishing a quantitative relationship between the magnitude of the response and the dose responsible for inducing such a response. This frequently entails extrapolation from relatively high levels of exposure used in experimental studies to significantly lower levels that are characteristic of the ambient environment. Of central importance to this effort is the shape of the dose–response function below the experimentally observable range and therefore the range of inference, i.e., whether it is linear or curvilinear and, if so, in what direction (Figure 7.2). Also, in the case of carcinogens, there is the question of whether the curve intercepts zero response only at zero dose or, as suggested by the premise of a physiologic reserve, there is a population threshold below which such effects are not manifested. To address this issue and others related to risk assessment, a wide range of decision support methods have been proposed to extend the plausible range of biologically based inference in extrapolations used in dose–response assessment (Wilson *et al.* 1995).

The exposure assessment element of risk assessment attempts to identify potential or completed exposure pathways resulting in contact between the agent and at-risk populations. It should also include a demographic analysis of at-risk populations, describing properties and characteristics of the populations

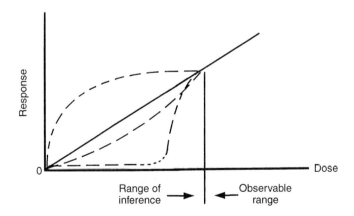

Figure 7.2 Dose–response curve, with emphasis on the shape of the dose–response function below the experimentally observable range and therefore the range of inference.

that might potentiate or mitigate concern regarding potential exposures. Finally, exposure assessment should include a description of the magnitude, duration, and frequency of exposure.

Ideally, in the risk characterization component, hazard identification, dose–response assessment, and exposure assessment are integrated into a qualitative and/or quantitative assessment characterizing the probability of adverse health effects in an exposed population. In practice, however, probabilistic projections are limited to carcinogens based on the application of low-dose extrapolation models. For this reason there is growing interest in the application of benchmark-dose approaches for noncarcinogens. While risk assessment is used in a wide range of environmental health and related programs, it has also been the subject of significant critical review and comment (Office of Technology Assessment (OTA) 1993; National Research Council 1994).

Both qualitative and quantitative risk assessment have utility in public health practice. ATSDR uses such estimates as well as professional judgement, external, peer, and public review in its decisions. This approach has been effective in addressing the many challenges posed in risk assessment, including the use of animal data, extrapolation from high to low dose, and exposure assessment, as well as in characterizing uncertainties in the risk assessment as part of the risk characterization process. In its review of risk assessment, the NAS (National Research Council 1994) recommended that the current approach to assessing risk should be retained. Indeed, much criticism directed at risk assessment, while justified, focuses not so much on the risk assessment paradigm itself, but rather on how the risk assessment data are applied and interpreted, specifically the tendency to characterize risk using single-point estimates, often conveying an artificial sense of precision that is not justified by the available data.

The NAS also recommended that a tiered, multilevel, or iterative approach to risk assessment be pursued. In this approach, relatively inexpensive techniques are used to screen for chemicals and/or exposure scenarios that do not pose a significant health risk. At successive levels of this hierarchy, increasingly rigorous methods are employed that rely less on default assumptions and more on measured exposure parameters of the at-risk population. In this way, resources could be applied to those issues that are likely to pose the most significant hazards to human health based on more realistic estimates of actual health risk than highly conservative, overstated estimates of risk that often result from the use of defaults.

In parallel with the NAS study, the OTA (1993) issued a report on health risk assessment research. Key conclusions of that report included a need for the greater integration of research across federal agencies and a tighter linkage in the continuum of applied and basic research, risk assessment, and risk management. Risk management represents the culmination of the regulatory decision-making process that incorporates all information on benefits versus risks of exposure in specific situations (Figure 7.3). Enhancing such linkages would allow for improvements in the default-based approach to risk assessment by improving

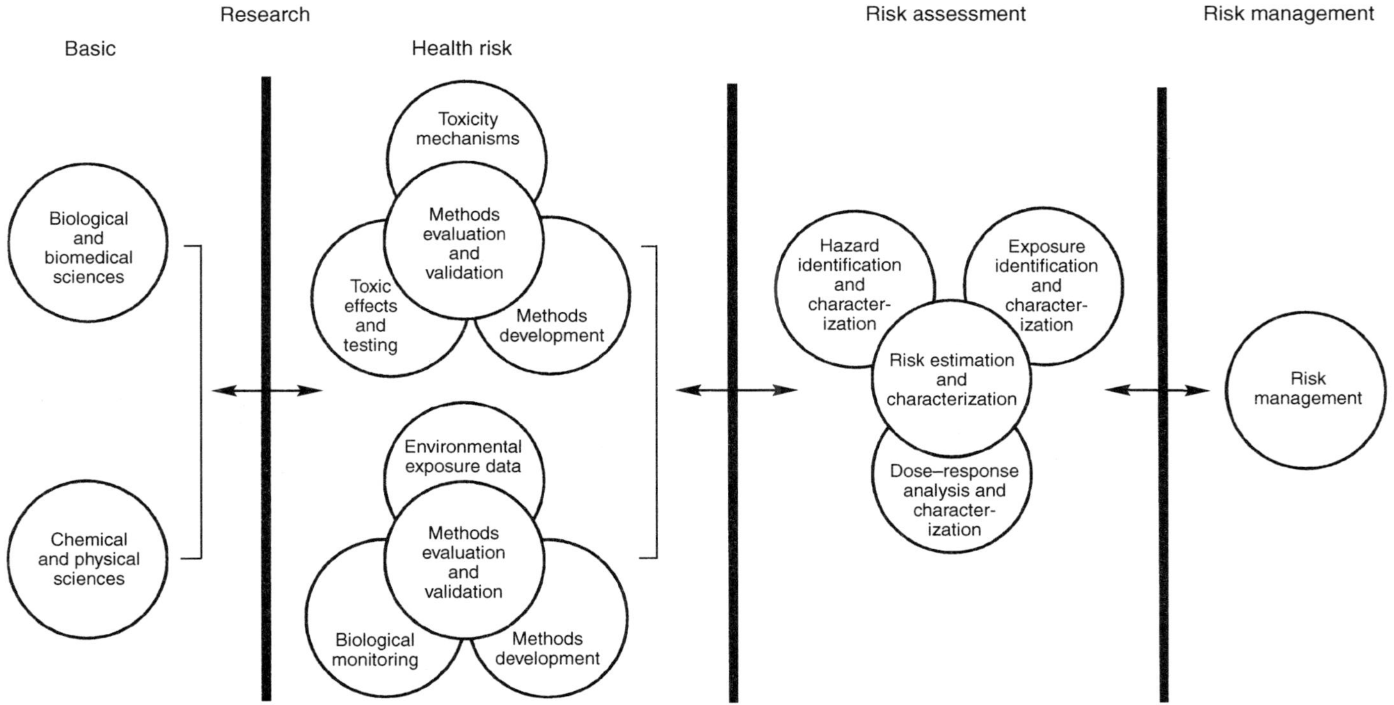

Figure 7.3 Linking research on health risks to decision making (OTA 1993).

both models and data used in analysis, and thus strengthen the technical basis for decision making in risk management.

Such an approach is consistent with the use of risk analysis as defined by the Council on Environmental Quality (CEQ 1989). Within this framework, risk analysis serves as an organizing construct based on sound biomedical and other scientific judgement, as well as risk assessment, to define plausible exposure ranges of concern, rather than single numerical values that may be misinterpreted or overinterpreted. In this approach, biomedical opinion, host factors, molecular epidemiology, and actual exposure conditions are considered critically important in evaluating the significance of environmental exposures to hazardous substances (ATSDR 1993). As applied by ATSDR in public health practice, risk analysis is a multidimensional endeavor, encompassing the components of risk assessment, biomedical judgement, risk communication, and risk management (Figure 7.4).

Within this framework of risk analysis, ATSDR may rely on one or more elements of risk assessment in pursuing public health practice. Risk assessment methods are employed for screening, organizing information, setting priorities, defining research needs, and making decisions. For example, the agency develops both minimal risk levels (MRLs) and environmental media evaluation guides (EMEGs) and has proposed a framework for defining significant human exposure levels (SHELs) on a site-specific basis (Figure 7.5). In subsequent

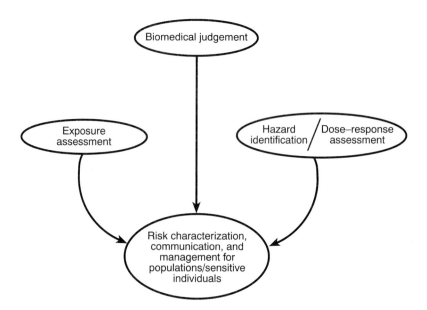

Figure 7.4 The elements of health risk analysis with emphasis on biomedical judgement.

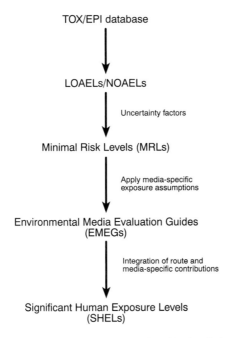

Figure 7.5 Development of the Agency for Toxic Substances and Disease Registry substance-specific guidelines for noncancer endpoints. TOX = toxicology; EPI = epidemiology; LOAELs = lowest-observed-adverse-effect levels; NOAELs = no-observed-adverse-effect levels.

sections of this chapter, the interrelationships among and scientific bases for these and other risk assessment approaches are presented as they relate to public health practice at ATSDR.

7.2 RISK ASSESSMENT AS AN ORGANIZING CONSTRUCT FOR SCIENTIFIC INFORMATION

As mentioned previously, risk assessment includes an evaluation of information on the hazardous properties of substances. It can be viewed as a systematic approach to organizing and analyzing scientific knowledge and information on hazardous substances that may pose risks under specific conditions (National Research Council 1994). Furthermore, as it

'provides a highly organized profile of the current state of knowledge of particular issues and systematically elucidates scientific uncertainties, it can provide valuable guidance to research scientists regarding the type of data that can most effectively improve understanding' (National Research Council 1994).

The role of risk assessment as an organizing construct for scientific information is exemplified by the agency's activities involving: (1) chemical-specific assessment; (2) multimedia assessment; (3) the substance-specific applied research program (SSARP); and (4) the chemical mixtures research program.

7.2.1   CHEMICAL-SPECIFIC ASSESSMENT

The agency is mandated to develop toxicological profiles for hazardous substances most commonly found at Superfund sites and that pose the most significant threats to human health as determined by ATSDR and the Environmental Protection Agency (EPA). Each profile includes an examination, summary, and interpretation of available toxicologic information and epidemiologic evaluations of a hazardous substance and provides estimates of human exposure levels for that substance associated with short- and long-term health effects. Based on these evaluations, the agency derives MRLs for inhalation and oral exposure.

An MRL is defined as 'an estimate of the daily human exposure to a substance that is likely to be without an appreciable risk of adverse, noncancer effects over a specified duration of exposure' (Pohl and Abadin 1995; ATSDR 1996a). For each exposure route, MRLs can be derived for acute ($\leqslant 14$ days), intermediate (15–364 days), and chronic ($\geqslant 365$ days) exposure durations. These health guidance values are derived from the most sensitive endpoint for each exposure route and duration, using either the highest no-observed-adverse-effect level or the lowest-observed-adverse-effect level. Generally, uncertainty factors (Ufs) of 10 are used in calculating MRLs. However, other Ufs may be applied under certain circumstances (Pohl and Abadin 1995). To date, ATSDR has derived 242 MRLs. MRLs are not intended for use in regulatory action but as a guide for toxicity assessment. Within the agency, MRLs are frequently used in developing EMEGs as described in the section below, 'Risk assessment as a decision support tool'.

7.2.2   MULTIMEDIA ASSESSMENT

As previously discussed, MRLs are derived for single chemicals by single exposure routes and specific exposure durations. However, actual environmental exposures are far more complex. Therefore, the agency is developing a multimedia framework that will enable health assessors to identify SHELs on a site-specific basis (Mumtaz et al. 1995). This decision support framework is based on the report of a working group convened by the International Programme on Chemical Safety (IPCS) and chaired by ATSDR (IPCS 1993). It provides a decision tree linking health guidance values, such as MRLs, with site-specific data, using tools/models that support biologically based inference and extrapolation. These tools/models include physiologically based pharmacokinetic (PBPK) modeling, structure–activity relationships, benchmark-dose modeling, and determination of interaction factors (Ifs) for chemical mixtures based on the weight of evidence for chemical interaction (Mumtaz *et al.* 1995). Therefore, this conceptual framework can be used to develop exposure and toxicological information for hazardous substances with limited databases.

Briefly, the derivation of SHELs involves three steps: estimation of total integrated exposure (TE), estimation of total tolerable level (TL), and comparison of the two values. The TE is estimated by combining t1e contributions of each medium and each route of exposure to the total chemical exposure. The TL is calculated by partitioning MRLs or other health guidance values according to the relative contribution of each route. If TE exceeds TL, there may be a cause for concern, i.e., a SHEL has been identified. The methodology for deriving SHELs is described in detail by Mumtaz *et al.* (1995). This multimedia framework is not intended to be used to develop a list of numbers (SHELs); rather, it is intended to lead to a qualitative evaluation of a specific waste site to determine whether a significant human exposure has occurred.

7.2.3 PRIORITY AND HYPOTHESIS-DRIVEN RESEARCH

7.2.3.1 Substance-Specific Applied Research Program

Section 104(i)(5) of CERCLA directs the administrator of ATSDR (in consultation with the administrator of the EPA and agencies and programs of the Public Health Service) to assess whether adequate information on the health effects of priority hazardous substances found at National Priorities List (NPL) sites is available. Where adequate information is not available, ATSDR, in cooperation with the National Toxicology Program (NTP), is required to ensure the initiation of research designed to fill these gaps.

The major purpose of this SSARP is to address the substance-specific informational needs of the public and the scientific community. This program will provide ATSDR with the necessary information for improving the database that is used to conduct public health assessments of populations living in the vicinity of hazardous waste sites. This program will also provide data that can be generalized to other substances or areas of science thus creating a scientific base for addressing a broader range of data needs. The program's purposes are more fully described in ATSDR (1989).

The SSARP was initiated on October 17, 1991. At that time, a list of priority data needs for 38 priority hazardous substances was announced in the Federal Register (ATSDR 1991). The list was subsequently revised based on public comments and published in final form on November 16, 1992 (ATSDR 1992a). It contains 117 priority data needs associated with the 38 substances. Priority data needs for 12 additional priority hazardous substances were recently announced in the Federal Register (ATSDR 1996b).

In Section 104(i)(5)(D), CERCLA states that it is the sense of Congress that the costs for conducting this research program be borne by the manufacturers and processors of the hazardous substances under the Toxic Substances Control Act of 1976 (TSCA) and by registrants under the Federal Insecticide, Fungicide, and Rodenticide Act of 1972 (FIFRA), or by cost recovery from responsible parties under CERCLA. To execute this statutory intent, ATSDR developed a plan whereby parts of the SSARP are supported via regulatory mechanisms (TSCA/FIFRA), private sector voluntarism, and through the direct use of CERCLA funds. This is described in the section below, 'Toward the future: linkage issues'.

7.2.3.2 Chemical mixtures research program

Section 104 of CERCLA, as amended, established the requirement that ATSDR develop methods for determining the health effects of chemical mixtures at hazardous waste sites. It is recognized that populations are typically exposed to multiple chemicals at these sites. Furthermore, the behavior of chemicals in mixtures might differ greatly from that observed for single chemicals. Interactions of the mixture components might alter toxicity through mechanisms such as synergism, antagonism, potentiation, inhibition, and masking.

The agency has developed a strategy to address this complex issue that consists of three elements: (1) a trend analysis for commonly occurring combinations of chemicals; (2) an assessment of available information on potential interactions; and (3) an applied research program for chemical mixtures (Johnson and De Rosa 1995).

During 1994, ATSDR initiated development of methods to assess the health effects of chemical mixtures found at hazardous waste sites. Towards this end, ATSDR made awards (cooperative agreements and contracts) to implement this research program as described in the section below, 'Toward the future: linkage issues'.

7.3 RISK ASSESSMENT AS A DECISION SUPPORT TOOL

Another role risk assessment may assume is that of a decision support tool. This role is evident in ATSDR's program activities. For example, health risk assessment is used as a screening tool in the agency's petition process (see below) and the application of media-specific comparison values. It is also used as a tool for setting priorities, e.g., in the agency's listing of hazardous substances and site ranking.

7.3.1 SCREENING

7.3.1.1 The petition process

As a public health agency, ATSDR is mandated to assess the effects of hazardous waste on the health of populations living in the vicinity of hazardous waste sites. ATSDR may be petitioned by members of an affected community, other interested groups, or government officials to conduct a public health assessment. In response to the request, the agency team, including scientists and physicians, will visit the site and interview members of the community. Subsequently, the team will evaluate all site information and present its results to the ATSDR petition committee, which will determine if any action needs to be taken. If the committee decides that a public health action is necessary, it may recommend a public health assessment, a public health advisory, a health consultation, or community environmental health education, as appropriate. Since 1987, ATSDR has received more than 350 petitions from the public. About 60% are from community residents. The remainder are from companies, attorneys, legislators, and grassroots activists.

7.3.1.2 Media-specific comparison values

As the first step in conducting a public health assessment, ATSDR assesses what contaminants are of concern at the site (ATSDR 1992b; Abouelnasr, personal communication, 1996). The agency develops media-specific comparison values that are used to select these contaminants of concern. These values represent concentrations in air, water, or soil below which no human health effects are expected to occur. The comparison values serve as a screening tool to determine if the contaminants need to be considered in the public health assessment of the site. They are not intended to be regulatory values or cleanup standards.

Two kinds of comparison values are available, i.e., those based on noncancer effects and those based on cancer effects. The EMEGs for noncancer effects are derived using ATSDR's inhalation and oral MRLs for specific exposure durations, and assumptions about exposure. In the absence of MRLs, other values, such as the EPA's reference dose or reference concentration, may be utilized (ATSDR 1992b).

The cancer risk evaluation guides (CREGs) for cancer effects are derived using the EPA's inhalation unit risk and oral cancer slope factors for inhalation and oral exposures, respectively. These values are used for calculating the concentration in an environmental medium corresponding to a given cancer risk. For ATSDR, the acceptable risk is set at 10^{-6}. In cases where inhalation unit risk or oral slope factors are not available for known or suspected carcinogens, ATSDR will list the substance as a contaminant of concern if it is detected at the site (Abouelnasr, personal communication, 1996).

7.3.2 PRIORITY SETTING

7.3.2.1 Listing of hazardous substances

ATSDR is mandated, in collaboration with the EPA, to develop a priority list of hazardous substances found at NPL sites. These substances are determined to

Table 7.1 Ten top-ranked hazardous substances

Rank	Substance
1	Lead
2	Arsenic
3	Mercury, metallic
4	Vinyl chloride
5	Benzene
6	Cadmium
7	Polychlorinated biphenyls
8	Chloroform
9	Benzo[a]pyrene
10	Trichloroethylene

Source: Johnson and De Rosa (1995).

pose a human health risk based on: (1) known or suspected human toxicity; (2) frequency of occurrence at NPL sites or other facilities; and (3) the potential for human exposure to the substance. The list is used by the agency in setting priorities for its activities, e.g., the development of toxicological profiles and the implementation of the SSARP. Every year, the agency reexamines the hazardous substances found at NPL sites using its database, HazDat, to ensure that the most hazardous substances are included in the list. To date, 275 substances have been included in the priority list (ATSDR 1994b). The 10 top-ranked substances are shown in Table 7.1.

7.3.2.2 Site ranking

The final step in developing a public health assessment includes identifying levels of public health hazard posed by the site and recommending appropriate follow-up activities based on the health implications of the site. In selecting the appropriate health hazard category, health assessors need to consider all site-specific information. Important factors to be considered include, but are not limited to, completed or potential exposure pathways, potential for multiple exposures, contaminant interactions, presence of sensitive subpopulations, and community health concerns (ATSDR 1992b).

The agency has established five public health hazard categories (ATSDR 1992b).

1. Category A, 'urgent public health hazard', is used for sites that pose an urgent public health hazard due to short-term exposures to hazardous substances.
2. Category B, 'public health hazard', is used for sites that pose a public

Table 7.2 Sites by Agency for Toxic Substances and Disease Registry hazard category in public health assessments (PHAs)

Category	All PHAs (n = 1714)	FY 1992 (n = 233)	FY 1993–1994 (n = 132)
Urgent public health hazard	1%	2%	5%
Public health hazard	22%	35%	49%
Indeterminate public health hazard	66%	41%	34%
No apparent public health hazard	7%	20%	10%
No public health hazard	3%	2%	2%
Unclassified	1%	0%	0%

Data from Johnson and De Rosa (1995).
FY, fiscal year.

health hazard as the result of long-term exposures to hazardous substances.

3. Category C, 'indeterminate public health hazard', is used for sites with incomplete information.
4. Category D, 'no apparent public health hazard', is used for sites where human exposure to contaminated media is occurring or has occurred in the past, but the exposure is below the level of concern.
5. Category E, 'no public health hazard', is used for sites that do not pose a public health hazard.

For the period 1992–1994, Table 7.2 shows the ATSDR hazard categories assigned to uncontrolled waste sites that underwent public health assessment.

7.3.2.3 Health Activities Recommendation Panel

ATSDR has established an agency-wide panel, the Health Activities Recommendation Panel (HARP), to determine the appropriate follow-up activities to the agency's public health assessments, public health advisories, health consultations, and other health-related activities. The three major types of follow-up actions are environmental health education, health studies, and substance-specific applied research. The choice of action is linked to the degree of hazard a site poses (ATSDR 1992b).

7.4 THE ROLE OF PUBLIC HEALTH PRACTICE IN GUIDING THE EVOLUTION OF RISK ASSESSMENT

As previously discussed, risk assessment is a means of spanning the gulf between available data and the conclusions regarding the nature of human health hazards associated with hazardous substances in the environment. As such, it can be considered as a prospective assessment or working hypothesis regarding health hazards for hypothetical populations. The degree to which such prospective assessments for hypothetical populations apply generically to the full range of populations is a key question. ATSDR attempts to address this issue directly through its activities at hazardous waste sites.

In pursuing its mandated interventions, e.g., health surveillance, ATSDR is in a unique position to assess the degree to which the prospective hypothetical risk assessments have either underestimated or overestimated the human health risk associated with a particular exposure scenario. Thus, if the risk assessment is viewed as a working hypothesis to characterize risks, the results of health interventions are the means by which to refine, modify, or accept this working hypothesis. By virtue of this feedback, the biological basis of risk assessment models can be greatly enhanced by new epidemiology and toxicology data, and by new insights into mechanisms of action. These data are directly relevant to the uncertainties and gaps in knowledge in the risk assessment process as they reflect genetic susceptibility, dose–response relationships, the concordance of effects between animal and human populations, and special groups at risk (Figure 7.6).

In sharing its findings with a range of stakeholders in the public health community, ATSDR attempts to present and harmonize its findings in the context of all relevant risk assessments, reviews, and decisions. This includes presentation of all assumptions and limitations as well as the intended purpose(s) of such assessments. ATSDR believes it is important to have interaction between the agency and the affected community. Therefore, an agency priority is to involve the community during the planning and assessment activities for a site to address its site-specific health concerns. Furthermore, the agency considers education activities for the community to be equally important.

7.5 TOWARD THE FUTURE: LINKAGE ISSUES

As discussed above, a major objective of the ATSDR SSARP is to supply information to improve the database for conducting public health assessments. A limiting factor in most, if not all, risk assessments is incomplete information. The research capabilities of both federal and private organizations are useful, directly or indirectly, in addressing the need for more data to strengthen the biological inference and extrapolation processes used in risk assessment.

In response to a request by Congress and as noted previously, the OTA evaluated the nature, organization, and management of federally supported research on health risk assessment. It concluded that:

'In times of limited, even declining Federal budgets, research linkages among the Federal Government, industry, and universities are critical for

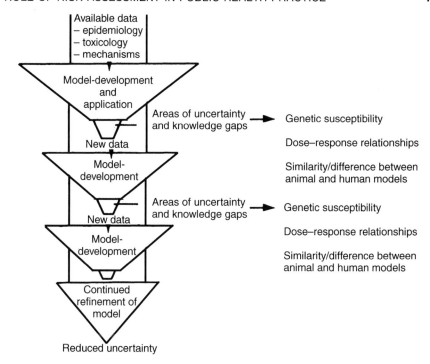

Figure 7.6 Development of biologically based risk assessment models. Available data are used to develop models. Areas of uncertainty and data gaps are identified that, when addressed, will lead to refined models with reduced uncertainty. Adapted from Thigpen *et al.* (1993).

advancing health risk assessment research. These linkages could be important for at least three reasons: they infuse more resources into the field; they bring together researchers with different backgrounds, expertise, and interests; and they increase the trust between the public and private sectors' (OTA 1993).

ATSDR, other federal agencies, and nongovernment organizations routinely interact to foster research collaborations. In addition to support via regulatory mechanisms (TSCA/FIFRA), private sector voluntarism, and through the direct use of CERCLA funds, data needs are being addressed through an interagency agreement with the National Toxicology Program (NTP), by ATSDR's Great Lakes Human Health Effects Research Program, and other agency programs. To date, a total of 63 research needs associated with 38 ATSDR priority hazardous substances (including 15 polycyclic aromatic hydrocarbons (PAHs)) are being addressed via these mechanisms (ATSDR 1996c).

7.5.1   FEDERAL AGENCIES

7.5.1.1   Tri-agency Superfund Applied Research Committee

A Tri-Agency Superfund Applied Research Committee (TASARC), comprising scientists from the ATSDR, NTP, and EPA, has been set up to: (1) provide a forum to discuss substance-specific research activities being carried out via TSCA/FIFRA, private sector voluntarism, or Superfund; (2) coordinate knowledge of research activities to avoid duplication of research being conducted in other programs and under other authorities; and (3) maintain a scheduled forum that provides an overall review of the ATSDR Superfund Applied Research Program.

TASARC has met seven times since the initiation of the SSARP. It has guided identification of data needs for referral to the EPA and associated development of enforceable consent agreements under TSCA and FIFRA. In addition, it has referred data needs to the appropriate private sector organizations that will conduct voluntary research. Furthermore, TASARC has become a forum for other federal agencies to bring forth their research agendas. For example, it has coordinated research efforts on hazardous pollutants with the Office of Air and Radiation, EPA. TASARC has developed testing guidelines for immunotoxicity, and it has endorsed the use of decision support methodologies such as PBPK modeling and benchmark-dose modeling. TASARC is also addressing issues related to the speciation of metals found at hazardous waste sites.

7.5.1.2   TSCA/FIFRA

In developing and implementing the SSARP, ATSDR, NTP, and EPA have established procedures to identify priority data needs of mutual interest to a number of federal programs. These data needs are being addressed through toxicologic testing under TSCA and FIFRA (ATSDR 1996c). This portion of the research will be conducted according to established TSCA and FIFRA procedures and guidelines. Generally, this testing will address more than one federal program's need. Priority data needs being addressed by EPA rule making are shown in Table 7.3.

7.5.1.3   National Toxicology Program

ATSDR maintains an interagency agreement (IAG) with NTP to conduct toxicologic testing of substances identified at NPL sites. Recently, the IAG was modified to include toxicity studies of ATSDR's priority hazardous substances via application of structure–activity relationship (SAR) techniques and PBPK modeling. The IAG will also support toxicology studies on the use of monitoring of receptor-mediated processes for screening ATSDR's priority hazardous substances for noncancer effects.

Table 7.3 Priority data needs being addressed by Environmental Protection Agency rule making

Substance	Priority data need	TSCA/FIFRA
Mercury	Dose–response data in animals for chronic-duration oral exposure	TSCA
	Immunotoxicology battery of tests via oral exposure	
Benzene	Dose-response data in animals for acute- and intermediate-duration oral exposure. The subchronic study should include an extended reproductive organ histopathology	TSCA
	Neurotoxicology battery of tests via oral exposure	TSCA
Chromium	Dose–response data in animals for acute-duration to chromium(VI) and (III), and for intermediate-duration to chromium (VI) via oral exposure	TSCA
	Multigeneration reproductive toxicity study via oral exposure to chromium (III) and (VI)	TSCA
	Immunotoxicology battery of tests following oral exposure to chromium (III) and (VI)	TSCA
Cyanide	Dose–response data in animals for acute- and intermediate-duration exposures via inhalation. The subacute study should include extended reproductive organ histopathology and evaluation of neurobehavioral and neuropathological endpoints	TSCA (inhalation study)
	2-Species developmental toxicity study via oral exposure	TSCA
	Evaluation of the environmental fate of cyanide in soil	TSCA
Beryllium	Dose–response data in animals for acute- and intermediate-duration inhalation exposures. The subchronic study should include extended reproductive organ histopathologyTSCA	TSCA
	2-species developmental toxicity study via inhalation exposure	TSCA
	Environmental fate in air; factors affecting bioavailability in air	TSCA
	Immunotoxicology battery of tests following oral exposure	TSCA (inhalation study)
Toluene	Dose–response data in animals for acute- and intermediate-duration oral exposures. The study should include an extended histopathological evaluation of the immune system	TSCA
	Comparative toxicokinetic studies (characterization of absorption, distribution, and excretion via oral exposure)	TSCA
DEHP	Comparative toxicokinetic studies (studies designed to examine how primates metabolize and distribute DEHP as compared with rodents via oral exposure)	TSCA
Chloroethane	Dose–response data in animals for acute- and intermediate-duration oral exposures. The subchronic study should include an evaluation of immune and nervous system tissues (and behavior, demeanor) and extended reproductive organ histopathology	TSCA
		(EPA will address only the immune system requirement of this priority data need)

FIFRA, Federal Insecticide, Fungicide, and Rodenticide Act of 1972; TSCA, Toxic Substances Control Act of 1976; DEHP, Di[2-ethylhexyl] phthalate.

In addition, ATSDR has referred candidate mixtures from its chemical mixtures research program to NTP for testing using the decision support methods described earlier.

7.5.2 ACADEMIC INSTITUTIONS/HEALTH DEPARTMENTS/ CONTRACTORS

7.5.2.1 CERCLA-funded research (Minority Health Professions Foundation Research Program)

In 1992, ATSDR announced a $4 million cooperative agreement program with the Minority Health Professions Foundation (MHPF) to support substance-specific investigations. This cooperative venture has been supported by CERCLA funds of about $4 million annually since the initiation of the program. Up to $4 million is available to continue the program through 1998.

Currently, nine priority data needs for 21 priority hazardous substances (including 15 PAHs) in the SSARP are being addressed by the MHPF institutions through this program (Table 7.4) (ATSDR 1996c).

A not-for-profit 501(c)(3) organization, the MHPF comprises 11 minority health professions schools. Its primary mission is to research the persistent health problems that disproportionately affect poor and minority citizens. The purposes of the ATSDR–MHPF cooperative agreement are to: (1) initiate research to address ATSDR-identified data needs for priority hazardous substances, and (2) enhance existing disciplinary capacities to conduct research in environmental health at MHPF member institutions.

7.5.2.2 Great Lakes Human Health Effects Research Program

Some of the priority data needs identified in the SSARP have been independently identified as research needs through the ATSDR Great Lakes Human Health Effects Research Program, a separate research program. To date, 12 priority data needs for 19 priority hazardous substances (including 15 PAHs) identified in the SSARP are being addressed through this program (Table 7.5) (ATSDR 1996c).

The Great Lakes Critical Programs Act of 1990 mandated the EPA, in consultation with ATSDR, to prepare a report that assesses the adverse effects of pollutants in the Great Lakes system on the health of individuals in the Great Lakes states. In support of this directive, ATSDR received funds to conduct research. The ATSDR-supported research projects focus on at-risk populations to further define the human health consequences of exposure to persistent toxic substances in the Great Lakes basin.

In 1992, ATSDR initiated a $2 million grant program to conduct research on the human health effects due to consumption of contaminated fish from the Great Lakes region. Recipients of grants include state universities and health

Table 7.4 Priority data needs being addressed by Minority Health Professions Foundation institutions

Substance	Priority data need	Institution
Lead	Mechanistic studies on the neurotoxic effects of lead	Florida A & M University Texas Southern University
	Exposure levels in humans living near hazardous waste sites and other populations, such as exposed workers	The King/Drew Medical Center of the Charles R. Drew University of Medicine and Science Morehouse School of Medicine
Mercury	Multigeneration reproductive toxicity study via oral exposure	Tuskegee University
Benzene	2-species developmental toxicity study via oral exposure	Xavier University
PAHs	Dose–response data in animals for intermediate-duration oral exposures. The subchronic study should include extended reproductive organ histopathology and immunopathology	Meharry Medical College
	Dose–response data in animals for acute- and intermediate-duration inhalation exposures. The subchronic study should include extended reproductive organ histopathology and immunopathology	Meharry Medical College
Trichloroethylene	Neurotoxicology battery of tests via oral exposure	Texas Southern University
Toluene	Neurotoxicology battery of tests via oral exposure	Texas Southern University
Zinc	Dose–response data in animals for acute- and intermediate-duration oral exposures. The subchronic study should include an extended histopathological evaluation of the immunologic and neurological systems	Xavier University Tuskegee University

PAHs, polycyclic aromatic hydrocarbons.

Table 7.5 Priority data needs being addressed by the Agency for Toxic Substances and Disease Registry Great Lakes Human Health Effects Research Program

Substance	Priority data need	Institution
Lead	Exposure levels in humans living near hazardous waste sites and other populations, such as exposed workers	State University of New York at Buffalo State University of New York at Oswego Michigan State University University of Wisconsin—Superior New York State Health Department University of Illinois at Chicago University of Illinois at Urbana-Champaign Wisconsin Department of Health and Social Services
Mercury	Multigeneration reproductive toxicity study via oral exposure	State University of New York at Oswego University of Illinois at Chicago
	Exposure levels in humans living near hazardous waste sites and other populations, such as exposed workers	State University of New York at Buffalo State University of New York at Oswego Michigan State University University of Wisconsin—Superior New York State Health Department University of Illinois at Chicago University of Illinois at Urbana-Champaign Wisconsin Department of Health and Social Services
PCBs	Potential candidate for subregistry of exposed persons	Wisconsin Department of Health and Social Services University of Wisconsin—Superior
	Dose–response data in animals for acute- and intermediate-duration oral exposures	
	Epidemiological studies on the health effects of PCBs (special emphasis endpoints include: immunotoxicity, gastrointestinal toxicity, liver, kidney, thyroid toxicity, reproductive and developmental toxicity)	State University of New York at Buffalo State University of New York at Oswego University of Wisconsin—Superior University of Illinois at Chicago University of Illinois at Urbana-Champaign

	Research need	Institution
	Exposure levels in humans living near hazardous waste sites and other populations, such as exposed workers	State University of New York at Buffalo State University of New York at Oswego Michigan State University University of Wisconsin—Superior New York State Health Department University of Illinois at Chicago University of Illinois at Urbana- Champaign Wisconsin Department of Health and Social Services Wisconsin Department of Health and Social Services
PAHs	Epidemiological studies on the health effects of PAHs (special emphasis endpoints include cancer, dermal, hemolymphatic, and hepatic)	Wisconsin Department of Health and Social Services
	Exposure levels in humans living near hazardous waste sites and other populations, such as exposed workers	State University of New York at Buffalo State University of New York at Oswego Michigan State University University of Illinois at Chicago Wisconsin Department of Health and Social Services
DDT	Epidemiological studies on the health effects of DDT, DDD, and DDE (special emphasis end points include: immunotoxicity, reproductive and developmental toxicity)	
	Exposure levels in humans living near hazardous waste sites and other populations, such as exposed workers	State University of New York at Buffalo State University of New York at Oswego Michigan State University University of Illinois at Chicago Wisconsin Department of Health and Social Services Wisconsin Department of Health and Social Services
	Potential candidate for subregistry of exposed persons	

PAHs, polycyclic aromatic hydrocarbons; PCBs, polychlorinated biphenyls.

departments. Funds of $3 million (for 1993 and 1994), $4 million (for 1995), and $2.2 million (for 1996) were allocated to support 10 research projects. Funding for 1997 is anticipated to be at least at the 1996 level.

In addition, the ATSDR chemical mixtures research program funded three universities (cooperative agreements) and a private company (contract). The research projects include development of alternative toxicologic approaches to evaluate chemical mixtures, *in vitro* testing coupled with limited *in vivo* testing, and development of computational methods.

7.5.3 PRIVATE SECTOR VOLUNTARISM

As part of the SSARP, ATSDR established a set of procedures for conducting voluntary research (ATSDR 1992c,d). Private sector organizations are encouraged to volunteer to conduct research to address these specific priority data needs. ATSDR intends to enter into only voluntary research projects that lead to high-quality scientific work and for which data can be shared with the public. Therefore, all study protocols and results are evaluated by ATSDR's peer reviewers. The agency will accept the research projects upon the recommendations of the peer reviewers.

ATSDR has been pursuing voluntary research interests with three private sector organizations: the General Electric Company (GE), the Halogenated Solvents Industry Alliance (HSIA), and the Chemical Manufacturers Association (CMA). Through the voluntary research efforts of these organizations, data needs for two classes of substances (polychlorinated biphenyl compounds (PCBs) and volatile organic compounds) are being addressed (ATSDR 1996c). Priority data needs being addressed by voluntary research are shown in Table 7.6. To date, three memoranda of understanding (MOUs) have been signed by ATSDR and the interested parties. Recently, the agency received a letter of intent from the Independent Zinc Alloyers Association volunteering to conduct research to address ATSDR's priority data needs for zinc.

In 1995, ATSDR entered into an MOU with GE. This marked the first time a private sector organization had volunteered to conduct research to address data needs identified in ATSDR's SSARP. The MOU with GE covers studies on PCBs, including an assessment of the chronic toxicity and oncogenicity of Aroclors (commercial PCB mixtures) administered in a diet to rats, PCB congener analyses, and metabolite detection as a tool for determining naturally occurring, aerobic PCB biodegradation.

In addition, ATSDR signed an MOU with HSIA covering studies to address three ATSDR priority toxicity data needs for methylene chloride. The studies evaluated acute- and subchronic-duration oral exposures, and developmental toxicity via oral exposure. The data were obtained using PBPK modeling.

Currently, HSIA and ATSDR continue to discuss voluntary research efforts for trichloroethylene and tetrachloroethylene.

Recently, ATSDR signed an MOU with CMA covering studies to address two

Table 7.6 Priority data needs being addressed by voluntary research

Substance	Priority data need	Firm
PCBs[a]	Chronic toxicity and oncogenicity via oral exposure	General Electric Company
	Aerobic PCB biodegradation in sediment	General Electric Company
	PCB congener analysis	General Electric Company
Methylene chloride[b]	Dose–response data in animals for acute- and intermediate-duration oral exposure. The subchronic study should include extended reproductive organ histopathology, neuropathology and demeanor, and immunopathology	Halogenated Solvents Industry Alliance
Vinyl chloride	2-species developmental toxicity study via the oral route	Halogenated Solvents Industry Alliance
	Combined 2-generation reproductive and developmental toxicity study via the inhalation route	Chemical Manufacturers Association

[a]Not *priority* data needs.
[b]Data were obtained by physiologically based pharmacokinetic modeling. Voluntary research for trichloroethylene and tetrachloroethylene is under discussion with the Halogenated Solvents Industry Alliance.
PCBs, polychlorinated biphenyls.

ATSDR priority data needs for vinyl chloride, i.e., reproductive and developmental toxicity via inhalation exposure.

7.5.4 PUBLIC INPUT

ATSDR is responsive to public opinion and concerns regarding its program activities. Towards that end, the agency invites the public to comment on its documents and procedures. For example, in developing its toxicological profiles, the agency announces the availability of these draft documents in the *Federal Register*, invites the public to comment on these drafts, and considers the public comments in preparing the final versions of these documents. Similarly, ATSDR was aware of concerns within some segments of the public regarding voluntary research conducted by companies with vested interests in the research. Therefore, ATSDR held two public meetings to discuss the development of its voluntary research procedures, and encouraged the public to comment on the draft procedures when they were announced in the *Federal Register*. Furthermore, the agency requests comments from the public when it announces the priority data needs for ATSDR's priority hazardous substances.

7.6 CONCLUSIONS

We have described the impact of risk assessment and risk analysis on the programs of ATSDR. As our comments indicate, ATSDR's preference is to follow the risk analysis paradigm developed by the NAS. ATSDR considers its public health assessments at uncontrolled hazardous waste sites (i.e., Superfund sites), and actions associated with the health assessment, to be consistent with a risk analysis approach, which emphasizes a tiered, multilevel, or iterative approach to health risk assessment. It is ATSDR's view of its public health mission under CERCLA and other federal environmental statutes that risk assessment by itself does not provide an adequate platform for determining community health interventions. This is because there are too many uncertainties in the human health effects and toxicological databases to permit easy application of risk assessment to determine public health interventions, e.g., to select those Superfund sites at which to conduct epidemiologic investigations.

The ATSDR experience with risk assessment, risk analysis, and health assessment has taught several lessons. It has demonstrated the value of setting priorities for hazardous substances and then identifying key data gaps for those substances. This perspective forces government to identify and defend its data priorities and it is hoped that the resultant research to fill data gaps will lead to improved risk assessments and health assessments. That, in turn, should improve the application of risk assessment to determine more cost-effective cleanup levels at Superfund sites and elsewhere. The data gap provisions of

CERCLA, as amended, have been underappreciated for their implications for improving site- and chemical-specific risk assessments.

Another lesson learned was the importance of linking human health databases with toxicologic databases. This kind of linkage, one in which human exposure data are compiled along with toxicological data, enhances the practice of site-specific health assessments and risk assessments. This enhancement occurs because health assessors are able to view information regarding potential exposure, ongoing exposure, and completed exposure pathways (including body burdens) in the context of effects and effect levels that have been demonstrated in experimental studies. Human exposure data for persons exposed to releases from Superfund sites are vital for determining public health interventions and their priority.

In closing, it is unfortunate that the Superfund statute, because of its controversial nature, has become a program focused almost exclusively on site cleanups and litigation. What is lost as a result of this narrow focus is an appreciation of the considerable body of improved scientific data, improved risk assessment methods, and positive impacts on community health that have resulted during the implementation of the Superfund statute and that have been funded by Superfund. As funds are redirected away from science and public health into litigation and cleanups, the public will ultimately be the poorer for this de-emphasis of health issues. Stagnant environmental and toxicological science and less clear understanding of impacts on human health will be the result.

ACKNOWLEDGMENTS

The authors thank Dr. Dennis Jones for reviewing this manuscript, Dr. John Wheeler for preparing Tables 7.3–7.6, Ms. Anne A. Olin for her editorial assistance, Ms. Mary Knox for her secretarial assistance, and members of the ATSDR Visual Information Center for preparing the graphics.

REFERENCES

Agency for Toxic Substances and Disease Registry (1989). Decision guide for identifying substance-specific data needs related to toxicological profiles. *Federal Register* **54**: 37618–37634.

Agency for Toxic Substances and Disease Registry (1991). Identification of priority data needs for 38 priority hazardous substances. *Federal Register* **56**: 52178–52185.

Agency for Toxic Substances and Disease Registry (1992a). Announcement of final priority data needs for 38 hazardous substances. *Federal Register* **57**: 54150–54159.

Agency for Toxic Substances and Disease Registry (1992b). *Public Health Assessment Guidance Manual.* US Department of Health and Human Services, Public Health Service, Washington, DC.

Agency for Toxic Substances and Disease Registry (1992c). Procedure for conducting voluntary research. *Federal Register* **57**: 4758–4761.

Agency for Toxic Substances and Disease Registry (1992d). Revised procedures for conducting voluntary research. *Federal Register* **57**: 54160–54163.

Agency for Toxic Substances and Disease Registry (1993). *Cancer Policy Framework*. US Department of Health and Human Services, Public Health Service, Washington, DC.

Agency for Toxic Substances and Disease Registry (1994a). *FY 1994 Agency Profile and Annual Report. October 1, 1993, to September 30, 1994.* US Department of Health and Human Services, Public Health Service, Washington, DC.

Agency for Toxic Substances and Disease Registry (1994b). Revised priority list of hazardous substances that will be the subject of toxicological profiles. *Federal Register* **59**: 9486–9487.

Agency for Toxic Substances and Disease Registry (1996a). Minimal Risk Levels for priority substances and guidance for derivation; Republication. *Federal Register* **61**: 33511–33520.

Agency for Toxic Substances and Disease Registry (1996b). Identification of priority data needs for 12 priority hazardous substances. *Federal Register* **61**: 14430–14438.

Agency for Toxic Substances and Disease Registry (1996c). Update on the status of the Superfund substance-specific applied research program. *Federal Register* **61**: 14420–14427.

Council on Environmental Quality (1989). *Risk Analysis: A Guide to Principles and Methods for Analyzing Health and Environmental Risks*. PB89-137772. National Technical Information Service, Washington, DC.

De Rosa CT, Choudhury H, Schoeny RS (1993). Information needs for risk assessment. In: *The Access and Use of Information Resources in Assessing Health Risks from Chemical Exposure*. US Environmental Protection Agency and Oak Ridge National Laboratory, Oak Ridge, TN.

International Programme on Chemical Safety (1993). The Derivation of Guidance Values for Health-based Exposure Limits. International Technical Report. United Nations Environment Programme, World Health Organization, Geneva.

Johnson BL (1995). ATSDR's information databases to support human health risk assessment of hazardous substances. *Toxicology Letters* **79**: 11–16.

Johnson BL, De Rosa CT (1995). Chemical mixtures released from hazardous waste sites: Implications for health risk assessment. *Toxicology* **105**: 145–156.

Mumtaz MM, Cibulas W, De Rosa CT (1995). An integrated framework to identify significant human exposures (SHELs). *Chemosphere* **31**: 2485–2498.

National Research Council (1983). *Risk Assessment in the Federal Government: Managing the Process*. National Academy Press, Washington, DC.

National Research Council (1994). *Science and Judgment in Risk Assessment*. National Academy Press, Washington, DC.

Office of Technology Assessment (1993). *Researching Health Risks*. US Congress, Washington, DC.

Pohl HR, Abadin HG (1995). Utilizing uncertainty factors in minimal risk levels derivation. *Regulatory Toxicology and Pharmacology* **22**: 180–188.

Thigpen K, Maloney DM (1993). Toxicity tests in animals: extrapolating to human risks. *Environmental Health Perspectives* **101**: 396–401.

Wilson JD, Cibulas W, De Rosa DT, Mumtaz MM, Murray E (eds) (1995). Decision support methodologies for human health risk assessment of toxic substances. Proceedings of the 1993 Decision Support Methodologies International Workshop. *Toxicology Letters* **79**: 1–312.

8

An Example of Cooperative Risk Assessment: Scenario Analysis for the Risk of Pine Shoot Beetle Outbreaks Resulting from the Movement of Pine Logs from Regulated Areas

ROBERT L. GRIFFIN
Animal and Plant Health Inspection Service, US Department of Agriculture, Riverdale, MD, USA

8.1 INTRODUCTION

The broad field of ecological risk analysis includes a set of unique challenges posed by exotic organisms that represent hazards because of the damage they can do to plant and animal resources. The Animal and Plant Health Inspection Service (APHIS) is the agency within the US Department of Agriculture (USDA) that is charged with managing the risks associated with exotic plant and animal pests that may be spread via human-assisted pathways, such as commodities moving in international commerce or domestic trade.

The success of APHIS regulatory efforts largely depends on the strength of risk analysis as the basis for appropriate actions. Another important factor is the degree of support APHIS receives from cooperators such as state regulatory authorities and the industries involved with implementation of or affected by regulatory programs. Frequently, these groups have different perspectives and information that may lead them to conclusions and decisions that differ from

This chapter is a US Government work and, as such, is in the public domain in the United States of America.

those identified by APHIS. Resolving such differences is not only desirable for rule making, but it is essential for implementing effective cooperative programs.

However, building consensus and developing support can be extremely difficult when the issues are controversial, highly publicized, complex, or strongly influenced by politics or biased interests. Under such circumstances, APHIS needs to emphasize the technical and scientific bases for decision making. This process brings risk analysis into the spotlight, and moves disputes into a more systematic and objective forum for resolution.

Risk analysis techniques that utilize expert groups (expert information approaches) have been used by APHIS to help assemble and evaluate relevant evidence as well as to develop consensus positions incorporating diverse interests (Kaplan 1992). Using expert information approaches to support a probabilistic risk assessment has proven especially effective for resolving controversial or complex issues involving a high degree of uncertainty.

The following discussions provide an overview of risk analysis in APHIS and briefly describe an example of the use of an expert information approach in combination with a probabilistic risk assessment method known as scenario analysis (Miller *et al.* 1993). The example is a cooperative assessment performed to evaluate possible regulatory strategies for managing pine shoot beetle (PSB), *Tomicus piniperda* (L), an exotic timber pest recently introduced into Michigan and surrounding states.

8.2 RISK ANALYSIS IN THE ANIMAL AND PLANT HEALTH INSPECTION SERVICE

Zero risk is clearly an unreasonable objective for regulatory agencies charged with protecting plant and animal health in an environment of accelerating global trade. A more realistic goal is to manage the risk to achieve the maximum degree of safety that can be provided within the available resources. Managing pest risk in a regulatory context means that APHIS is continually estimating and prioritizing risks to ensure that resources are devoted to managing the most important risks and that regulatory decisions lead to initiatives that provide an appropriate level of protection without undue burden to regulators, the regulated community, or the consumer.

Achieving this balance requires a strong commitment to analysis as the basis for decision making. Good decisions require information. Because perfect knowledge is not attainable, the state of our knowledge and the associated uncertainty are important considerations in decision making. Risk analysis provides a systematic approach to evaluating the evidence and uncertainty that forms the basis for decisions.

APHIS defines risk analysis to include risk assessment, risk management, and risk communication. Decision making may be viewed as a separate activity that uses the results of risk analysis in the context of other influential variables.

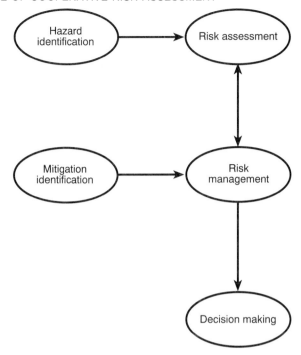

Figure 8.1   Paradigm for risk analysis.

Figure 8.1 shows the relationship between risk assessment, risk management, and decision making. Several specific elements are embedded in each of these components, but the important features to note are the distinction between decision making and technical analysis, and the interrelationship between risk assessment and risk management.

The products of risk assessment are conclusions about the likelihood and magnitude of the risk, as well as the uncertainty associated with these estimates. From these products a judgement is made concerning the acceptability of the risk and whether remedial actions are required. The process advances to risk management if the risk is deemed to be worthy of mitigation. In this case, options are identified and evaluated at three levels: efficacy, feasibility, and impacts (Figure 8.2).

It has been argued that considering the feasibility and impact of risk manage-ment options falls outside the scope of risk analysis because these elements of the analysis deal with the risk of making a decision rather than the risk presented by the primary hazard, e.g., a pest. This view suggests that the technical, e.g., biological analyses extend only to the point where efficacy is considered and the degree to which risk is reduced is the only criterion for prioritizing options.

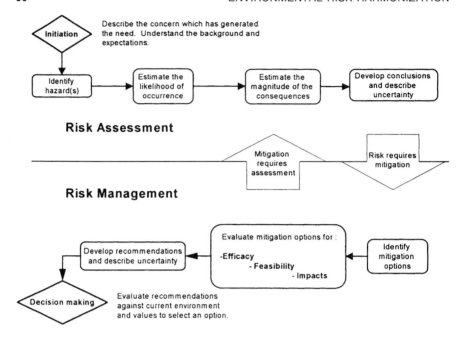

Risk Assessment

Risk Management

Figure 8.2   A process overview for pest and disease risk analysis.

This argument is largely academic because, in the final analysis, the risk manager will need as complete a picture as possible. This picture will have to include feasibility and impacts. Whatever the view taken, the products of risk management are recommendations and discussions of associated uncertainties that should be considered by those who make decisions.

8.3   THE ROLE OF THE ANIMAL AND PLANT HEALTH INSPECTION SERVICE IN HARMONIZATION

Because the theme of discussions here involves harmonization, it is appropriate to note that APHIS plays a key role in the harmonization of risk assessment at several levels. Not only is the agency concerned with harmonization across federal agencies and between federal and state regulatory bodies, but also at the international level.

The General Agreement on Tariffs and Trade (GATT) and the North American Free Trade Agreement (NAFTA) both contain risk analysis obligations for the United States under their respective Sanitary and Phytosanitary (SPS) Agreements (NAFTA 1993a; World Trade Organization 1994). International and regional standards for risk analysis have been developed and adopted by the organizations responsible for implementing each agreement. APHIS has been very active and influential in developing, understanding, and implementing these standards.

The development of these standards is significant for harmonization between countries, but it is also important for states. Unjustified inconsistencies between states or between the states and the federal government can provide the basis for challenges under GATT and NAFTA (NAFTA 1993b; World Trade Organization 1994). These challenges can have significant negative repercussions in the United States.

Adjusting our regulatory systems to accommodate concerns for compliance with our trade obligations has added another level of complexity to an already complex web of influences associated with these processes. On the other hand, it has also increased our sensitivity to trade issues, resulting in closer relationships with the regulated industries and new priorities given to management options that are not only justified by science, but are also as fair as possible to the private sector. The impact of this awareness can be seen in the changes in the handling of numerous risk issues. A pest risk assessment that has been developed cooperatively with federal and state agencies is used here to illustrate the successful application of an innovative approach to harmonization through cooperation in the resolution of an issue identified by industry.

8.4 THE PINE SHOOT BEETLE

PSB is a forest pest that was recently introduced into the United States from Europe (Haack and Kucera 1993). It is considered a serious pest, affecting pine trees and some other species of commercially valuable evergreens. The pest has become established in parts of Michigan and six other states in the Great Lakes region. The pathways for introduction are believed to be pallets, dunnage, and crating associated with imported cargo. This conclusion is supported by the observation that all outbreaks appear to have started near ports (APHIS Advisory Group on Pine Shoot Beetle, personal communication).

APHIS responded to the establishment of PSB by immediately conducting surveys and establishing quarantines for all areas where the pest was found. These quarantines severely restricted movement of unmanufactured wood products, specifically raw logs from these areas.

Because logging is an important industry in many parts of Michigan, the logging industry and the state became concerned about the restrictiveness of the quarantines and the serious impact of these regulations on local economies. This led to questions concerning the risk and the need to move from broad-based emergency quarantine measures to a more sophisticated management program for regulating affected areas.

Late in 1994, APHIS engaged in discussions with the Michigan Department of Agriculture (MDA) concerning proposals offered by MDA for regulating logs

moving from PSB-infested areas to non-infested areas. MDA suggested a regulatory program that involved significantly less stringent requirements than the
quarantine, arguing that the more rigorous program implemented by APHIS
was unfairly impacting the logging industry and was not justified by the risk.

APHIS has the regulatory responsibility for interstate movement of the
regulated material, including logs, Christmas trees, bark, wood chips, and
greenery (wreaths and garlands). MDA has the responsibility for regulating the
intrastate movement of the same commodities. Both parties recognized that a
consistent (harmonized) approach between key state and federal agencies was
essential, but that agreement on the appropriate regulatory measures would
first require agreement on the level of risk, specifically for log movement.

Several experts were identified including representatives of APHIS, MDA, the
Forest Service, a forestry expert from an uninfested state, and a PSB researcher.
The group assembled at APHIS headquarters for several days of discussion.
Relevant information was provided in advance and experts were encouraged to
bring any additional information that they considered useful. Experts were
directed to focus on the evidence and to strive for consensus based on their
interpretation of the evidence. Risk analysts from APHIS set the format and
facilitated the discussions.

The primary objectives of the expert discussions were: (1) agreement on the
scenario(s) and events within the scenarios that lead to pest establishment, and
(2) sound estimates representing appropriate probabilistic distributions for
each of the events within the scenarios.

The experts agreed on four scenarios: A, B, C, and D (Figure 8.3). Each
scenario represents a different pathway, and each point within a scenario
represents an independent event. Adding a seasonal component to the analysis
results in separate scenarios for each of five 'biological' seasons corresponding
to unique phases of the insect's life cycle and representing distinctly different
activities throughout the year. The result is 20 scenarios (four pathways × five
seasons). Table 8.1 shows how the data were organized and summarized.

Scenarios A and B relate to the movement and storage of unprocessed logs.
Scenario A is specific to the establishment of the pest as the result of escape in
transit. Scenario B addresses the likelihood of establishment due to introduc-
tions from mills and other storage and processing sites. Scenarios C and D
focus on the wood products. Scenario C involves logs that are processed into
lumber. Scenario D is for logs that are chipped. These are the two primary end
products for logs from the areas in question.

The steps involved in analyzing each pathway are illustrated in Figures 8.4
and 8.5. Table 8.1 summarizes the probability estimates developed by the
expert group for each event and season scenario. Each probability distribution
is described by a point estimate that defines what experts believed to be the
most likely value (the mode). Surrounding the point estimate are a low
(minimum) and high (maximum) that describes the experts' uncertainty
around the point estimate. Experts were encouraged to estimate conservatively

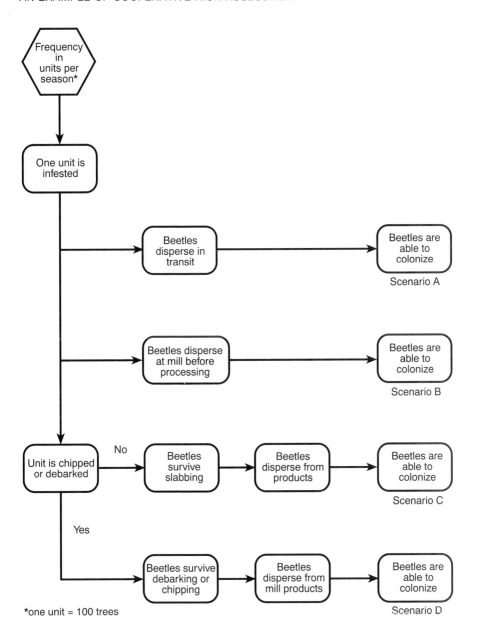

Figure 8.3 Combined scenarios for new outbreaks of pine shoot beetles due to the movement of logs.

Table 8.1 Combined probability estimates—all scenarios for pine shoot beetle/logs

	SUMMER Point	SUMMER Low	FALL Point	FALL High	FALL Low	WINTER Point	WINTER High	WINTER Low	EARLY SPRING Point	EARLY SPRING High	EARLY SPRING Low	LATE SPRING Point	LATE SPRING High
AF	225		225			225			112.5			112.5	
A1	0	0.01	0.45	0.7	0.05	0.5	0.8	0.25	0.6	0.9	0.5	0.2	0.4
A2		0.0002	0.02	0.05	0.0001	0.0075	0.03	0.05	0.125	0.25	0.05	0.125	0.25
A3		1E-06	1E-05	0.0001	1E-06	1E-05	0.0001	1E-05	0.0001	0.001	1E-06	1E-05	0.0001
Point product	2.0E-05		2.0E-05			8.4E-06			8.4E-04			2.8E-05	
BF	225		225			225			112.5			112.5	
B1	0	0.01	0.45	0.7	0.05	0.5	0.8	0.25	0.6	0.9	0.5	0.2	0.4
B2		1E-05	5E-05	0.0001	0.0001	0.0005	0.001	0.05	1.0E-01	0.3	0.5	0.1	0.3
B3		1E-06	1E-05	0.0001	1E-06	1E-05	0.0001	1E-05	0.0001	0.001	1E-06	1E-05	0.0001
Point product	5.1E-08		5.1E-08			5.6E-07			6.8E-04			2.3E-05	
CF	225		225			225			112.5			112.5	
C1	0	0.01	0.45	0.7	0.05	0.5	0.8	0.25	0.6	0.9	0.05	0.2	0.4
C2			0.5			0.5			0.5			0.5	
C3		0.90	0.95	0.99	0.9	0.95	0.99	0.9	0.95	.099	.09	0.95	0.99
C4		0.1	0.5	0.9	0.1	0.5	0.9	0.3	0.8	0.95	0.3	0.8	0.95
C5		0.0001	0.00225	0.05	0.0001	0.00225	0.05	0.001	0.0225	0.5	0.0001	0.00225	0.05
Point product	5.4E-02		5.4E-02			6.0E-02			5.8E-01			1.9E-02	
DF	225		225			225			112.5			112.5	
D1	0	0.01	0.45	0.7	0.05	0.5	0.8	0.25	0.6	0.9	0.05	0.2	0.4
D2			0.5			0.5			0.5			0.5	
D3		0.002	0.01	0.05	0.002	0.01	0.05	0.005	0.02	0.075	0.002	0.01	0.05
D4		0.001	0.0025	0.005	0.001	0.0025	0.005	0.005	0.0175	0.05	0.005	0.0175	0.05
D5		1E-06	1E-05	0.0001	1E-06	1E-05	0.0001	1E-05	0.0001	0.001	1E-06	1E-05	0.0001
Product point	1.3E-08		1.3E-08			1.4E-08			1.2E-06			2.0E-08	

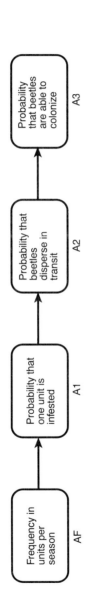

Frequency in units per season

AF → Probability that one unit is infested — A1 → Probability that beetles disperse in transit — A2 → Probability that beetles are able to colonize — A3

Combined probability estimates and point products for scenario A:

Event	Summer Point	Summer Low	Summer High	Fall Point	Fall Low	Fall High	Winter Point	Winter Low	Winter High	Early spring Point	Early spring Low	Early spring High	Late spring Point	Late spring Low	Late spring High
AF	225			225			225			112.5			112.5		
A1	0	0.01		0.45	0.05	0.7	0.5	0.25	0.8	0.6	0.05	0.9	0.2	0.05	0.4
A2	0.0002			0.02	0.0001	0.05	0.0075	0.05	0.03	0.125	0.05	0.25	0.125	0.05	0.25
A3	1.0E-06			1.0E-05	1.0E-06	0.0001	1.0E-05	1.0E-05	0.0001	0.0001	1.0E-06	0.001	1.0E-05	1.0E-06	0.0001
Point product				2.0E-05			8.4E-06			8.4E-04			2.8E-05		

Figure 8.4 Scenario A: dispersal in transit.

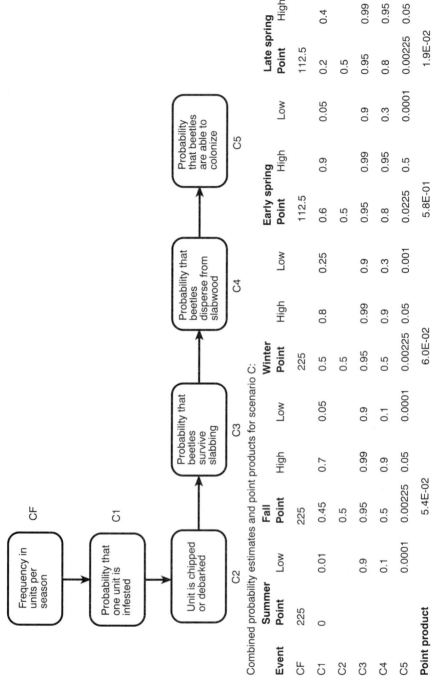

Figure 8.5 Scenario C: dispersal from slabwood

to ensure that the actual probability lies within the area of the curve defined by the estimates. A point estimate alone was used when the evidence indicated a very high degree of certainty. Estimates and calculations of probability stopped when any event eliminated the pest risk.

The results were interpreted a number of ways. The simplest method involved multiplying the point estimates to develop a 'most likely' estimate for each scenario. This was particularly useful for comparison purposes and to rank the scenarios according to their likelihoods (Table 8.2). Using this method immediately revealed scenario C as a much greater concern than the other scenarios.

Point estimates, however, do not describe the uncertainty associated with the estimates or the sum. To incorporate uncertainty, the distributions around each point estimate were generated and multiplied together. This was done with computer simulation using specially designed software, @Risk, to represent, graphically, the distributions for each event and to calculate the product of all events for each scenario (PC Windows).

Two types of distribution curves were generated using Monte Carlo simulation (Latin hypercube sampling) and 3000–9000 iterations (trials with random numbers). One type of curve is roughly 'bell-shaped', demonstrating the distribution of probability across the range of values defined by the

Table 8.2 Subscenario ranking by risk based on mode values

Product of point estimate	SCENARIO
1.3×10^{-8}	Fall dispersal from chips and bark at the mill site
1.4×10^{-8}	Winter dispersal from chips and bark at the mill site
2.0×10^{-8}	Late spring dispersal from chips and bark at the mill site
5.1×10^{-8}	Fall dispersal from logs at the mill before processing
5.6×10^{-7}	Winter dispersal from logs at the mill before processing
1.2×10^{-6}	Early spring dispersal from chips and bark at the mill site
8.4×10^{-6}	Winter dispersal from logs in transit
2.0×10^{-5}	Fall dispersal from logs in transit
2.3×10^{-5}	Late spring dispersal from logs at the mill before processing
2.8×10^{-5}	Late spring dispersal from logs in transit
6.7×10^{-4}	Early spring dispersal from logs at the mill before processing
8.4×10^{-4}	Early spring dispersal from logs in transit
1.9×10^{-2}	Late spring dispersal from slabwood and rough-cut lumber at the mill site
5.4×10^{-2}	Fall dispersal from slabwood and rough-cut lumber at the mill site
6.0×10^{-2}	Winter dispersal from slabwood and rough-cut lumber at the mill site
5.7×10^{-1}	Early spring dispersal from slabwood and rough-cut lumber at the mill site

Scenario A–Fall

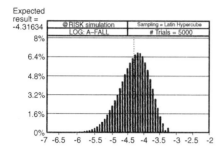

Scenario A–Winter

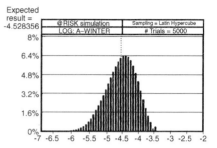

Scenario A–Early spring

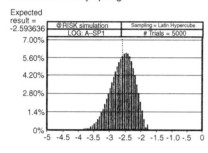

Scenario A–Late spring

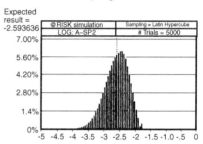

Scenario A–All Seasons

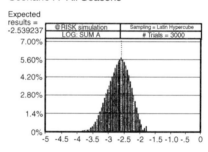

Figure 8.6 Probability distribution curves: scenario A.

experts. Figures 8.6 and 8.7 compare the curves for scenarios A and C; note that the distributions have been plotted on a log scale to make them easier to read.

Another curve that can be generated using the same results shows the cumulative probability from 0 to 100% (Figure 8.8). This curve is 's' shaped.

A tabular display of the results (Table 8.3) includes the mean, the mode, and

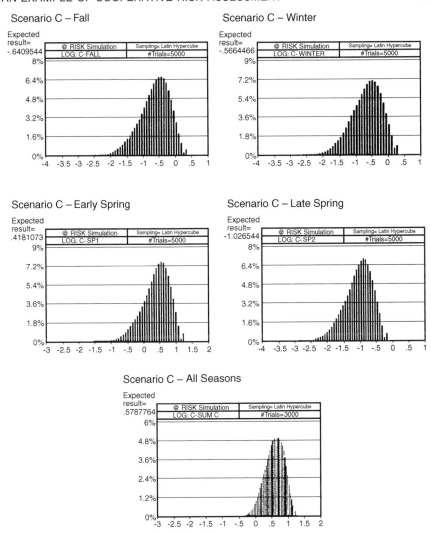

Figure 8.7   Probability distribution curves: scenario C.

the 95th percentile. Owing to the conservative estimates used by the experts, most of the distributions have very long 'tails' on the upper end of the probability distribution. The 'bottom line' in this example is the value calculated for estimated 'outbreaks per year' or 'years per outbreak'. To evaluate the potential effectiveness of a remedial action, these values can be compared with estimates of 'outbreaks per year' that are likely to occur without the application of risk management strategies.

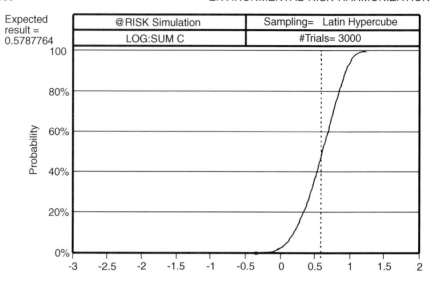

Figure 8.8   Scenario C: cumulative probability from log sum.

Table 8.3   Frequency of outbreaks by season and years per outbreak

Dispersal from:		Fall	Winter	Early spring	Late spring	New outbreaks year	Years/ outbreak
Scenario A transit	mean	7.9E-05	4.7E-05	3.4E-03	1.3E-04	3.7E-03	271
	mode	2.0E-05	8.4E-06	8.4E-04	2.8E-05	9.0E-04	1110
	95%	2.2E-04	1.4E-04	8.8E-03	3.3E-04	9.5E-03	105
Scenario B the mill	mean	1.7E-07	2.0E-06	3.7E-03	1.4E-04	3.8E-03	264
	mode	5.1E-08	5.6E-07	6.8E-04	2.3E-05	7.0E-04	1432
	95%	4.8E-07	5.3E-06	9.6E-03	3.6E-04	9.9E-03	101
Scenario C Slabwood	mean	3.6E-01	4.2E-01	3.7E + 00	1.4E-01	4.6E + 00	0.2
	mode	5.4E-02	6.0E-02	5.8E-01	1.9E-01	8.8E-01	1.1
	95%	1.0E + 00	1.2E + 00	9.2E + 00	3.7E-01	1.2E + 01	0.1
Scenario D mill by-products	mean	9.4E-08	1.1E-07	9.9E-06	2.3E-07	1.0E-05	97 290
	mode	1.3E-08	1.4E-08	1.2E-06	2.0E-08	1.2E-06	802 118
	95%	2.8E-07	3.4E-07	2.9E-05	7.1E-07	3.0E-05	33 078
All scenarios by season	mean	3.6E-01	4.2E-01	3.7E + 00	1.4E-01	4.6E + 00	0.2
	modfe	5.4E-02	6.0E-02	5.8E-01	1.9E-01	8.9E-01	1.1
	95%	1.0E + 00	1.2E + 00	9.2E + 00	3.7E-01	1.2E + 01	0.1

Another type of graphical summary (Table 8.4 and Figure 8.9) highlights the insignificance of scenarios A, B, and D, while showing very clearly that scenario C—the pathway for pest establishment via logs to be used for lumber—is where regulatory initiatives need to be focused. Further, it shows that about two-thirds of the risk is estimated to occur in one season. These strikingly clear results caused both APHIS and MDA to reconsider regulatory strategies.

Although a document was drafted by APHIS to describe and record the assessment process, it does not fully describe the process and the experience (Griffin and Miller 1994). Other highlights are included in the following.

1. The controversy associated with the issue was largely reduced to technical discussions leading to consensus on the important elements.
2. The use of expert information techniques including experts representing various sides of the issues resulted in diverse opinions being considered against diverse evidence to achieve a high degree of agreement on expected values and the range of uncertainty.
3. 'Buy-in' from dissimilar groups appeared to be increased substantially through participation on the expert panel and open consideration of all views and information. Cooperation significantly increased between key state and federal agencies as discussions proceeded.

Table 8.4 Percent risk by scenario and season using the mode value

Scenario	Fall	Winter	Early spring	Late spring	Sum
A: Transit	2.3E-03	9.5E-04	9.5E-02	3.2E-03	0.1
B: The mill	5.8E-06	6.3E-05	7.6E-02	2.5E-03	0.1
C: Slabwood	6.1	6.8	65.2	21.7	99.8
D: Mill by-products	1.5E-06	1.6E-06	1.4E-04	2.2E-06	00

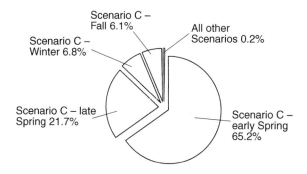

Figure 8.9 Comparative risk: based on mode values.

4. Critical areas and approaches for risk management were clarified and agreed upon.
5. Research needs and important information gaps were identified and prioritizing these needs was facilitated.
6. A high degree of credibility was associated with the product due to the level of expertise involved.

In sum, harmonization was successfully achieved through cooperation, objectivity, the use of innovative approaches, and a strong focus on the technical merits of judgements.

REFERENCES

Griffin RL, Miller CE (1994). Scenario analysis for the risk of pine shoot beetle outbreaks resulting from the movement of pine logs from regulated areas. Draft. Risk Analysis Systems, US Department of Agriculture, Animal and Plant Health Inspection Service, Riverdale, MD.

Haack B, Kucera D (1993). New introduction—common pine shoot beetle, *Tomicus piniperda* (L) Pest Alert NA-TP-05-93. US Department of Agriculture, Forest Service, Northeastern Area. Northeastern Forest Experiment Station, Radnor, Pennsylvania.

Kaplan S (1992). 'Expert information' versus 'expert opinions'. Another approach to the problem of eliciting/combining/using expert knowledge in PRA. *Reliability Engineering and System Safety* **35**: 61–72.

Miller L, McElvaine MD, McDowell RM, Ahl AS (1993). Developing a quantitative risk assessment process. Risk analysis, animal health and trade. *Revue Scientifique et Technique Office International des Epizooties* **12**: 1153–1164.

North American Free Trade Agreement between the Government of the United States of America, the Government of Canada, and the Government of the United Mexican States (1993a). *Agriculture and Sanitary and Phytosanitary Measures*, Vol. 1, Part 2, Chapter 7. Superintendent of Documents, US Government Printing Office, Pittsburgh, Pennsylvania.

North American Free Trade Agreement between the Government of the United States of America, the Government of Canada, and the Government of the United Mexican States (1993b). *Technical Barriers to Trade. Standards-related Measures*, Vol. 1, Part 3, Chapter 9. Superintendent of Documents, US Government Printing Office, Pittsburgh, Pennsylvania.

PC Windows version of @*Risk* for Lotus/Excel. Available from Palisade Corporation, Newfield, NY. Mention of companies or commercial products does not imply recommendations or endorsement by the US Department of Agriculture over others not mentioned. USDA neither guarantees nor warrants the standard of any product mentioned. Product names are mentioned solely to report factually on available data and to provide specific information.

World Trade Organization (1994). World Trade Agreement. Annex 1A: Agreements on Trade in Goods. Part 4: Agreement on the Application of Sanitary and Phytosanitary Measures. Geneva.

9

Past and Current Harmonization Efforts at the Federal Level

WILLIAM H. FARLAND
National Center for Environmental Assessment, US Environmental Protection Agency, Washington DC, USA

9.1 INTRODUCTION

The focus of this paper will be the efforts at harmonization of risk assessment that have occurred at the federal level during the last 10–15 years. The context for the discussion of risk harmonization is a statement in the 1983 National Research Council (NRC) report on risk assessment that the dominant analytic difficulty in making decisions based on risk assessment is pervasive uncertainty (NRC 1983). This reflects: (1) the variability in the characteristics of human and laboratory animal populations as well as environmental contaminant levels, and (2) the uncertainties in measuring adverse effects and environmental levels and extrapolating these to human populations of interest. Uncertainty in risk assessment makes harmonization very difficult because, while it may be easy to harmonize when all parties agree on the basic science of an issue, it is much more difficult to get agreement and harmonization in the face of uncertainty.

Further, uncertainty is important not only with regard to the types of risks and the probabilities and magnitudes of health effects, but also with respect to the social, political, and economic impacts of proposed efforts to manage a particular type of risk. While the issue of harmonization of risk assessment will be the main focus here, some consideration will be given to the relationship of risk assessment and risk characterization to regulatory decision making or risk management. It is important to understand this relationship as risk characterization is only one element considered in decision making, and it is often not the determining factor.

This chapter is a US Government work and, as such, is in the public domain in the United States of America.

9.2 FEDERAL INTERAGENCY EFFORTS

Federal agencies have made a number of efforts to harmonize risk assessment over the years. One of the first was the establishment of the Interagency Regulatory Liaison Group (IRLG) in the late 1970s. This group's charge was to bring federal agencies together to explore the possibility of developing uniform guidelines for health effects testing and assessment, and to a certain extent, risk management. IRLG work groups met for about 2 years to look at the state of the science for various endpoints used in risk assessment and to develop common risk assessment guidance. They made significant progress, especially with respect to assessing carcinogenic risk (IRLG 1979). However, when the new administration took office in 1981, the IRLG was abandoned because of the perceived impact it might have on regulatory activities in various federal agencies.

In the late 1980s, a group with a similar mission was appointed. It was called the Federal Coordinating Council for Science, Engineering, and Technology (FCCSET). FCCSET addressed a number of diverse risk issues, mainly through subcommittees assigned to these issues. The risk assessment subcommittee meetings provided a forum for debate and discussion of risk issues at some of the highest levels of federal government agencies, including the deputy administrator of the Environmental Protection Agency (EPA) and the deputy directors of other agencies. These meetings included debates about a broad range of risk issues from the development of risk assessment guidelines to the response to hurricanes and the Gulf War.

Three important products resulted from FCCSET activities. One was the development of consensus positions regarding some of the principles of risk assessment, positions that are reflected in the risk assessment principles proposed by a number of groups within the last few years. The second was a comparison between engineering risk analysis and health and ecological risk assessment that resulted in a publication which illuminated the relationships between the approaches of these two very separate disciplines (FCCSET 1992).

The third product was an effort to harmonize the cross-species scaling factor, an issue that had been very difficult to resolve because the differences between the EPA's and the Food and Drug Administration's (FDA) scaling factors were largely responsible for the inconsistencies between their respective risk assessment values, particularly for carcinogens. This effort resulted in agreement among the federal agencies to use an alternative approach, rather than scaling by either body weight or body surface area. This alternative scaling factor was body weight to the three-quarters power. This agreement was the result of a careful re-analysis of all the available data, and a consensus that this was a better value than that used by either the EPA or the FDA (US EPA 1992).

Presently, there is an effort by the National Science and Technology Council (NSTC) to address the principles to be used for risk assessment and cost–benefit analysis. There is also a Committee on Environmental and Natural Resources (CENR) that has a role similar to that of the FCCSET and the IRLG.

The risk assessment subcommittee of CENR is providing the vehicle for a significant amount of collaboration and interagency review of emerging risk assessment issues. It has the potential to play a very positive part in promoting risk harmonization.

9.3 RISK PUBLICATIONS

In addition to the outputs of these groups, there have been several landmark publications that have influenced federal risk harmonization efforts. *Risk Assessment in the Federal Government: Managing the Process*, published by the NRC in 1983, has had a profound effect on how federal agencies assess and manage risk (NRC 1983). This report presented a paradigm for risk assessment that focused on cancer risk assessment but, in fact, provided the impetus for discussion of noncancer risk assessment and ecological risk assessment as well.

In 1985, the Office of Science and Technology Policy (OSTP) published the results of an interagency effort addressing the principles of chemical carcinogen risk assessment in a report entitled *Chemical Carcinogens: Review of the Science and its Associated Principles* (US OSTP 1985). While the report provided guidelines in areas where consensus among agencies had not been reached, it had an impact on the EPA, FDA, and Occupational Safety and Health Administration cancer guidelines and/or policies. *Science and Judgment in Risk Assessment*, published in 1994 by the NRC as a response to the Clean Air Act Amendments of 1990, provided a revised look at the principles of risk assessment and recommendations for improvement (NRC 1994). Another result of the Clean Air Act amendments was the formation of the Commission on Risk Assessment and Risk Management (CRARM), which recently issued a draft report on a variety of issues related to risk assessment and risk management in the federal government, including risk harmonization, and included recommendations for change (CRARM 1996; Appendix B).

9.4 ENVIRONMENTAL PROTECTION AGENCY EFFORTS

The EPA has made a number of contributions to the federal harmonization effort that have affected the way that risk assessment is done at the federal, regional, and state levels (for examples see Table 9.1). The risk assessment guidelines that the EPA has published (US EPA 1989a), have had a very significant impact on the way that risk is assessed and managed. Recently, the EPA proposed new cancer guidelines in draft form and their potential impact is reflected in the 350 people who attended a Washington workshop devoted to a discussion of these guidelines (US EPA 1996).

Another EPA document, *The Exposure Factors Handbook*, provides default methodologies and the ranges of exposure factors that are used in Superfund

Table 9.1   Important Environmental Protection Agency
contributions to the federal harmonization effort

Risk assessment guidelines
Framework for ecological assessment
Exposure Factors Handbook
Science policy resolution
    toxicity equivalents for dioxin-like compounds
    scaling factor re-analysis
    species-specific tumor responses
Integrated Risk Information System

---

risk assessments and state risk assessments at contaminated sites (US EPA 1989c). In addition, science policy guidance related to toxicity equivalency factors, species-specific tumor responses, and fish advisory risk assessments are other examples of EPA efforts to promote harmonization (US EPA 1989b,d, 1991). Last, but not least, the Integrated Risk Information System (IRIS) provides toxicity values that promote consistency among governmental agencies (US EPA 1993).

9.5   RISK ASSESSMENT HARMONIZATION

All of this activity has led to a growing body of information at the federal level that has contributed to consensus about many desirable attributes of risk assessment. The documents cited above reflect the emergence of areas of agreement among risk assessment professionals regarding the appropriate execution of each of the components of the risk assessment process: hazard identification, dose–response assessment, exposure assessment, and risk characterization.

With respect to hazard identification, also known as hazard characterization, there is general agreement about the following: all relevant information should be presented and reviewed; the critical aspects of data quality should be highlighted, as quality is the most important issue with regard to scientific credibility; a weight-of-evidence approach should be utilized rather than basing conclusions only on positive studies; and research that would increase confidence in these conclusions should be identified. Once the types of research needs are identified more clearly, decisions can be made as to what research should be performed or information collected before regulatory decisions are made.

Turning to dose–response, there is consensus that valid data sets need to be presented, and that plausible models for high to low dose and interspecies extrapolation should be employed. Ideally, these models should provide a range of estimates for potency; however, if a single number is reported, it should be accompanied by explicit rationales for choices about data sets and models and defaults as well as information about the presence or absence of a scientific consensus about these models and defaults. The report should also show how dose–response relationships would be altered if alternative data sets, assumptions, and models are used.

With respect to exposure assessment, there is agreement that assessment methods, as well as measures of central tendency and high-end exposure, should be clearly described. In addition, the exposure evaluation should include descriptions of confidence in the data and the bases for the models, as well as measures of the validity of the models. Further, discussions of the influence of uncertainty and the sensitivity of the results to each the exposure parameter are critical for setting research agendas and for addressing risk management issues.

Finally, there is consensus that the risk characterization should present the major components of risk, including: quantitative estimates of each where appropriate; an integrative view of the evidence; key assumptions; and the rationale and extent of scientific consensus as well as the uncertainties involved in reaching the final conclusions. Further, information about the impacts of reasonable alternative assumptions on these conclusions should be provided.

While there is general consensus on numerous aspects of the risk assessment paradigm, several important issues remain. One is the relative importance of qualitative as compared with quantitative estimates as the bases for risk management decisions. Is there too much reliance on single numbers or quantitative estimates in making decisions? How can the scientific community provide risk managers with the most useful information about the strengths and weaknesses of the approaches that have been used?

9.6 RISK MANAGEMENT HARMONIZATION

While the focus of harmonization efforts to date have been on risk assessment, there have been some attempts at the federal level to deal with the harmonization of risk management as well. On a routine basis, the Office of Management and Budget (OMB) and the OSTP attempt to define the risk assessment and risk management principles that should guide federal agencies in their work. The OMB carries out reviews of regulations and very often the risk assessments that underlie the regulations, and provides detailed comments to the Office of the Science Advisor to the President. The Congress also attempts to influence harmonization through the development of legislation to mandate the way that risk assessments and, perhaps, risk management are performed. A recent example can be found in the risk provisions of the 1996 Food Quality Protection Act.

Risk management, as defined in Table 9.2, is based to some degree on risk assessment, but is heavily influenced by political, economic, and social information, much of which is also scientific. This latter type of scientific information is

Table 9.2   Risk management

Risk management can be considered as the complex synthesis of judgement and analysis that uses the results of a risk assessment, combined with political, economic, and social information to produce a decision about environmental action.
Risk management can also be considered as determining and accomplishing those actions that will reduce risk to the greatest degree, given any particular level of resources.
While individual risk management decisions may appear to be balancing risk reduction against resources, the system as a whole is designed to balance risk against risk.
It is designed to help us identify, and deal with, the worst and most controllable risks first.

---

often ignored because it is not the same science used in risk assessment. Within this framework, the role of risk assessment is to help the decision maker choose those actions that will reduce risk to the greatest degree within the limitations of available resources.

Although risk management decisions are generally narrowly focused, e.g., the risk of one chemical in one medium, and appear to be based on a balancing of risk reduction against resources, there is agreement that risk management decisions should be made in a broader context, e.g., multimedia and multi-chemical, and should be based on balancing risks against risks rather than risks against resources. One reason for this conclusion is that greatly increased knowledge is needed for society to address adequately the social, political, and economic issues that are critical to the effective balancing of risk reduction against the available resources. Owing to these gaps in knowledge, risk managers currently do not use a balancing approach weighing risks against resources but instead attempt to identify and deal with the worst and most controllable risks first.

The differences between risk assessment and risk management are illustrated in Table 9.3. Risk assessment activities are designed to identify, describe, and measure qualities and quantities that inform risk management. The risk manager then uses this information in conjunction with factors such as the social importance of the risk, the social acceptability of the risks, the economic impacts of risk reduction, and legislative mandates, when deciding on and implementing risk management approaches. There are fundamental differences in the way that risk assessment and risk management are performed but they are often confused so that risk assessment may be perceived as the source of a risk management decision, when, in fact, this decision may be based largely on social concerns, international issues, trade, public perception, or other non-risk considerations.

Table 9.3 Environmental risk: fields of analysis

Risk assessment	Risk management
Nature of effects	Social importance of risk
Potency of agent	De minimis or acceptable risk
Exposure	Reduce/not reduce risk
Population at risk	Stringency of reduction
average risk	Economics
high-end risk	Priority of concern
sensitive groups	Legislative mandates
Uncertainties of science	Legal issues
Uncertainties of analysis	Risk perception
Identify	*Evaluate*
Describe	*Decide*
Measure	*Implement*

A related risk management issue is comparative risk which requires application of both scientific and public policy analyses. If science is to be used validly in comparative risk studies, it must be based on health risk-to-health risk, ecological risk-to-ecological risk, or short-term risk-to-short-term risk comparisons. In addition, risks to be compared must be expressed in the same terms; all as actuarial risk values or all as forecasted risks. Comparisons of health versus ecological risks, voluntary versus involuntary risks, or actuarial versus forecasted risks decrease the rigor science can bring to the risk issues and move them into the arena of public policy. Unfortunately, this is not always recognized in reports of the results of increasingly common comparative risk analyses (e.g., California EPA 1994; Michigan Department of Natural Resources 1992).

Finally, risk communication is important to assure that the public understands the issues that have been raised. Providing clear lay language explanations of risk assessment findings and risk management choices, as well as the bases for these decisions, is a challenge that needs to be addressed. Hopefully in the future, risk assessment and risk management, including the way that conclusions have been reached, will be communicated much more effectively than at present.

REFERENCES

California Environmental Protection Agency (1994). Toward the 21st Century: Planning for the Protection of California's Environment. California Comparative Risk Project Final Report. Sacramento, CA.

Commission on Risk Assessment and Risk Management (1996). Risk Assessment and Risk Management in Regulatory Decision-Making (Draft). Washington, DC.

Federal Coordinating Council for Science, Engineering and Technology (FCCSET) Ad Hoc Working Group on Risk Assessment (1992). *Risk Assessment: A Survey of Characteristics, Applications, and Methods Used by Federal Agencies for Engineered Systems.* FCCSET, Washington, DC.

Interagency Regulatory Liaison Group, Work Group on Risk Assessment (1979). Scientific bases for identification of potential carcinogens and estimation of risks. *Journal of the National Cancer Institute* **63**: 241–268.

Michigan Department of Natural Resources (1992). *Michigan's Environment and Relative Risk.* Lansing, MI.

National Research Council (1983). *Risk Assessment in the Federal Government: Managing the Process.* National Academy Press, Washington, DC.

National Research Council (1994). *Science and Judgment in Risk Assessment.* National Academy Press, Washington, DC.

US Environmental Protection Agency (1989a). *Risk Assessment Guidance for Superfund,* Vol. 1. *Human Health Evaluation Manual* (Part A). Interim Final. EPA/540/1-89/002. Office of Emergency and Remedial Response, Washington, DC.

US Environmental Protection Agency (1989b). *Guidance Manual for Assessing Human Health Risks from Chemically Contaminated Fish and Shellfish.* EPA/503/8-89/002. Office of Marine and Estuarine Protection, Washington, DC.

US Environmental Protection Agency (1989c). *Exposure Factors Handbook.* EPA/600/8-89/043. Office of Health and Environmental Assessment, Washington, DC.

US Environmental Protection Agency, Risk Assessment Forum (1989d). *Interim Procedures for Estimating Risks Associated with Exposures to Mixtures of Chlorinated Dibenzo-p-Dioxins and Dibenzofurans (CDDs and CDFs) and 1989 Update.* EPA/625/3/89/016. Office of Health and Environmental Assessment, Washington, DC.

US Environmental Protection Agency, Risk Assessment Forum (1991). *Report of the EPA Peer Review Workshop on Alpha$_2$-globulin: Association with Renal Toxicity and Neoplasia in the Male Rat.* EPA/625/3-91/021. Washington, DC.

US Environmental Protection Agency (1992). Draft report: A cross-species scaling factor for carcinogen risk assessment based on equivalence of mg/kg$^{3/4}$/day. *Federal Register* **57**: 24152–24173.

US Environmental Protection Agency (1993). Integrated risk information system: announcement of availability of background paper. *Federal Register* **58**: 11490–11495.

US Environmental Protection Agency (1996). *Proposed Guidelines for Carcinogen Risk Assessment.* EPA/600/P-62/003C. Washington, DC.

US Office of Science and Technology Policy (1985). Chemical carcinogens: a review of the science and its associated principles. *Federal Register* **50**: 10371–10442.

10

State Approaches to Risk Assessment and Management: A Michigan Perspective

HAROLD HUMPHREY[1], JEFFREY A. CRUM[2] AND CHARLES P. CUBBAGE[3]

[1]*Michigan Department of Community Health, Lansing, MI, USA*

[2]*Environmental Response Division, Michigan Department of Environmental Quality, Lansing, MI, USA*

[3]*Michigan Department of Agriculture, Lansing, MI, USA.*

PART I HARMONIZATION FROM A PUBLIC HEALTH PERSPECTIVE *(Harold Humphrey)*

10.1 INTRODUCTION

It is very appropriate that the subject of harmonization is being addressed in the Great Lakes because eight states and Canada share responsibility for this valuable natural resource. This responsibility includes assessing and managing risks to human health and the environment from environmental contaminants. Although, it is clear that a number of such risks have been identified and reduced in recent years there is still a great deal left to do. Accomplishing these remaining goals will be greatly facilitated by increased coordination and harmonization among the Great Lakes governments.

While environmental contaminants in this region pose threats to both human health and the environment, this chapter will focus on public health as that is the mandate of the agency that the author represents. However, many of the same types of issues are relevant for environmental protection and it should also be recognized that data on risks to humans are often useful in understanding adverse effects in fish and wildlife.

This chapter is a US Government work and, as such, is in the public domain in the United States of America.

10.2 HARMONIZATION ISSUES

While discussions on harmonization often focus on the diversity of risk assessment and risk management approaches, one of the most significant barriers to risk harmonization is the lack of consistency among states and units within states with regard to the collection, evaluation, and availability of public health data. Examples of these data are found in vital records, disease registries, results of health and behavioral surveys, and measurements of contaminants in biological specimens, e.g., human or wildlife fluids and tissues.

Vital records generally provide information about causes of death and birth outcomes, including gross birth defects. While the information collected is well standardized nationally, it is generally not of the type and quality that is most useful for investigating relationships between environmental risks and human health. In particular, these data are limited because: (1) effects in humans from exposure to environmental contaminants are often more subtle than those recorded in vital records; (2) death may result from exposures many years in the past and so are not recognized contributors to mortality; and (3) deaths of interest may occur in small subpopulations whose records may be difficult to extract from the overall data.

Disease registries, such as cancer registries, have the potential to provide information that is more useful than vital records in relating exposures and effects as many of the individuals in these registries are living and can provide missing data about exposures. However, a serious problem with such registries is that different organizations may collect data about different diseases or collect data about the same diseases in different ways. As a result, combining information about populations in various locations is made more difficult and thus finding associations between environmental factors and health is more problematic.

Health surveys, such as the Behavioral Risk Factor Survey (Michigan Department of Public Health 1994), also have the potential to provide very useful data for assessing risk. Surveys can provide information about individual habits that can increase or decrease the susceptibility of individuals to environmental threats as well as individual activities that can lead to greater or lesser exposure to environmental contaminants. As with registry information, survey data often suffer from inconsistencies in the methods used for data collection and the types of information collected in different state and local jurisdictions.

10.3 HARMONIZATION EFFORTS

The problems described above were recognized many years ago by the Great Lakes states and Canada. In 1981, under the aegis of the International Joint Commission (IJC), representatives of each government met to discuss ways to increase harmonization in data collection (IJC 1981). The parties involved recognized that achieving such harmonization would be very beneficial for several reasons. One of the most important is that it would increase the potential study population from those living in a single state, e.g., 9 million people in Michigan, to those residing in the whole region, i.e., about 38 million people. With this much larger study population, epidemiological meta-analyses would have a much better chance of detecting relatively rare adverse impacts than would be possible using a single study of the population living in one state or province.

This IJC sponsored meeting also resulted in an exchange of information among representatives from several states that were developing cancer registries independently of one another. This, in turn, led to discussions of similarities and differences in approaches and to agreement among the participants from these states on a core of basic registry information that all would collect.

Following the meeting, information exchanges with other states were initiated and these led to the current situation where cancer incidence registries in the Great Lakes states are compatible with one another. Unfortunately, one serious problem persists, i.e., in some states the registry is protected by confidentiality laws and this can lead to the total or partial loss of vital information needed for scientific studies.

In addition to discussing cancer registries, state representatives exchanged information about other registries and vital information records. They discovered that each of the Great Lakes states and Canada collected at least one of the following types of information: health behaviors, birth outcomes, causes of death, and birth defect incidence. However, the types of information collected varied from state to state and, often, the same types of information were collected using different protocols in different states.

In light of the problems uncovered during these contacts, public health specialists in the Great Lakes region met at length in 1992 and discussed harmonization issues related to health information. The results of their deliberations were published in a report that described the existing situation, identified barriers to harmonization (such as inconsistent protocols), and contained recommendations for improvement (Institute for Environmental Studies 1992). One recommendation was that there should be a common minimum set of information collected. Another was that data should be collected across the basin using common protocols. A third was that laboratory analyses should be performed consistently in all laboratories with careful attention to quality assurance and quality control.

Following this meeting, steps were taken to implement these recommendations. One of the first steps was the agreement among four states to add the same questions to each state's Behavioral Risk Factor Survey so that they would all be collecting the same information on at least some behavioral variables. Additional steps became possible when the Agency for Toxic Substances and Disease Registry (ATSDR) established the Great Lakes Health Effects Research Program (ATSDR 1993) because one of the mandates of this program is to increase consistency of health data collection in the Great Lakes region.

In light of differences in fish consumption advice issued by the various Great Lakes states, one focus of this program is the collection of consistent data used to support such advisories. An example of this type of information is fish consumption data, including demographic information on the consumers. A unique feature of this effort is support for the collection of the fish consumption data by the program investigators, who are usually university scientists, rather than government agency personnel. Although this might appear to be a barrier to consistency as researchers often have somewhat different goals in their research and thus gather somewhat different data, researchers have met on this issue and have agreed to coordinate their surveys so that they are all asking at least some of the same questions. It is expected that these common data will be shared among project researchers so that basin-wide information will be available to all (ATSDR 1995).

Another type of information that the ATSDR Program is addressing is environmental monitoring data as harmonization of measurements of contaminant concentrations in fish is needed in addition to consistency in consumption data to facilitate common advisories among the states. Consistency in analyses of contaminants in humans is also important as these often provide the most reliable indicators of exposure.

Laboratory protocols are critical components of contaminant analyses and steps are underway to ensure quality assurance and quality control in the laboratories that are performing testing of human and environmental samples. As part of this process, researchers are developing a process by which laboratories can be evaluated to determine if they are operating within acceptable parameters and, if not, how they can change so that laboratories administered by different government units produce comparable results.

10.4 CONCLUSIONS

The eight Great Lakes states and Canada represent an important repository of health information about humans who live in this region, information that is potentially very useful in establishing links between environmental contaminants and human health. However, at present, the utility of these data is limited in some cases because: (1) different types of information have been collected in the various jurisdictions; (2) information has been collected using different protocols in different jurisdictions; (3) confidentiality reduces access to the information; and (4) the quality of information varies from jurisdiction to jurisdiction.

A number of steps have been taken in the Great Lakes during the past 15 years to reduce differences in data collection among agencies and to increase data quality. In recent years, commitments by the Great Lakes governments to improve environmental health and support from the ATSDR have been very helpful in facilitating the reduction of inconsistencies among jurisdictions. The governments in the Great Lakes region have learned a great deal about data harmonization during this process, including which processes have the greatest potential for encouraging harmonization and the types of data and protocols that are most easily harmonized. Thus, the governments in the Great Lakes region can serve as a valuable resource for other state and federal bodies that are interested in increased regional and national harmonization of risk assessment and risk management.

PART II MICHIGAN'S ENVIRONMENTAL CLEANUP REGULATIONS: INFLUENCES ON RISK ASSESSMENT *(Jeffrey A. Crum)*

10.5 INTRODUCTION

The use of risk assessment (RA) as a tool for guiding cleanup actions at sites of environmental contamination in Michigan has grown considerably since its first application in 1988. This is primarily due to the enactment of legislation which made RA a central remedial decision-making tool for administering and expediting environmental cleanups. In some cases, however, the legislatively devised structure of the cleanup program places limitations on the use of RA.

In addition, the standardized exposure assumptions and methods recommended in RA guidance from the Environmental Protection Agency (US EPA 1989, 1991a,b, 1996) often do not reflect the characteristics of contamination sites in Michigan. As a result, the methods and input assumptions developed for application on a nation-wide basis must be modified to characterize better the health risks posed by contamination sites in Michigan. Addressing this Michigan-specific need causes inconsistency with RA practices of the EPA and likely with other states.

This chapter, therefore, examines two principal factors that have created differences in the way that RA is used in Michigan compared with the EPA and other states. The first factor considers the influence of legislatively enacted environmental cleanup statutes on the use of RA in Michigan, while the second addresses how the Environmental Response Division (ERD) of the Michigan Department of Environmental Quality (DEQ) adapts the conventional RA equations and standardized assumptions in EPA guidance (US EPA 1989, 1991a,b) to develop generic groundwater and soil cleanup criteria for the protection of human health. Primary emphasis is given to characterizing human exposure scenarios that represent Michigan-specific conditions in developing the generic criteria.

The use of RA in Michigan's environmental remediation program originated in 1988 upon public approval of a bond issue. This action provided a significant funding source to the DEQ for conducting environmental cleanups. ERD soon identified approximately 2000 contaminated sites, prompting the need to develop an organized and effective method of prioritizing the threats posed by each site. The resultant priority process included development of health risk-based cleanup criteria.

The methods used to develop cleanup criteria for chemicals found in groundwater and soil have been described in the administrative rules for Act 307 of 1989, the Michigan Environmental Response Act (MERA). Three types of cleanup options, type A, B, and C, were specified to guide responsible parties (those liable for the contamination) in selecting the appropriate remedial action(s) for their site. Type A was defined as cleanup either to the method detection limit or to naturally occurring background concentrations, while a type B cleanup required reduction of contaminant concentrations in soils and groundwater to levels that would not pose unacceptable risks based on standardized exposure assumptions and acceptable risk levels. The type C cleanup option allowed responsible parties to conduct a site-specific RA using site-specific exposure assumptions to demonstrate the appropriateness of applying cleanup criteria other than type A or B criteria, while assuring protection of health and the environment.

During the period when the administrative rules were in effect (1989–1995), the majority of sites that were remediated used the type B cleanup option, although a small number of type C cleanups were approved. In addition, the type C site-specific cleanup option evolved to include standardized sets of criteria for industrial and commercial land uses as the exposure assumptions for type B cleanups were based on residential assumptions.

In June 1995, amendments to the Act 307 administrative rules (currently known as the Part 201 amendments) replaced the type A, B, and C terminology with several sets of land-use-based cleanup criteria. Under Part 201, the cleanup program changed from remediating nearly all sites to residential exposure-based criteria (i.e., type B) to a land-use-based framework with separate criteria for residential, industrial, and commercial land uses, with provision for development of recreational criteria.

The number of contaminated sites in Michigan currently exceeds 10 000, demonstrating a need for effective state priority setting within the cleanup program. Although conducting site-specific RAs remains an option for developing cleanup criteria, they are costly and time consuming to complete. Addressing this magnitude of sites requires a streamlined, straightforward approach that is simple, clear, and easy to implement. This inevitably leads to the application of RA as a rote methodology rather than a flexible tool that can be adapted to reflect site-specific conditions.

The present approach uses RA to develop chemical-specific cleanup criteria for each environmental medium (e.g., air, soil, groundwater) and each exposure pathway. These criteria serve as benchmarks for determining if significant human health risks may exist at a site and, if so, their priority for remediation. Through this approach, industry, the regulated community, and the general public are informed of the process used to identify and prioritize sites posing the greatest risks.

10.6 LEGISLATIVE IMPACTS

The application of health RA methods is strongly influenced by both the legislation governing cleanup and the ensuing risk management decisions necessary for implementing the law. The impact of the legislative process on both exposure assessment and toxicity assessment is apparent in the current environmental cleanup regulation; Part 201 of the Natural Resources and Environmental Protection Act, Act 451, as amended. With respect to exposure assessment, Part 201 requires the use of 'reasonable' exposure assumptions and 'relevant' exposure pathways in developing 'generic' land-use-based cleanup criteria.

For characterizing chemical toxicity, the cleanup regulations require that toxicity assessments for carcinogenic chemicals use the 95% upper confidence limit of the slope of the calculated dose–response curve. The target cancer risk was also changed in the regulation from 1 additional cancer in 1 000 000 exposed individuals to 1 in 100 000. For assessment of noncarcinogenic health effects, Part 201 requires using a Hazard Quotient (HQ) of 1. An HQ of 1 corresponds to a daily exposure level (or intake) that is equal to the reference dose (RfD). Furthermore, the regulation requires application of specialized, but scientifically supported approaches for certain contaminants through a provision stipulating the use of biologically based models, such as the Integrated Exposure Uptake Biokinetic (IEUBK) model for lead.

While certain legislative provisions in Part 201 direct the manner in which RA is used for establishing cleanup criteria, others were written to facilitate consistency with other Michigan and federal regulations. Thus, through legislation, one RA approach is adopted over another. The requirement that risk-based cleanup criteria be consistent with other environmental regulations promotes harmonization of regulatory standards although divergences in the RA methods remain. For example, if a state drinking water standard (i.e., maximum contaminant level (MCL)) has been promulgated for a chemical pursuant to Act 399 SDWA, it serves as the groundwater cleanup criterion regardless of the value derived using Part 201 RA methods.

10.7 RISK ASSESSMENT AND MANAGEMENT IMPACTS

After it is determined that a particular chemical is not affected by other regulations, RA is used to develop Part 201 generic land use-based criteria. A number of factors can affect the development of chemical-specific cleanup criteria (Figure 10.1) and can result in differences in RAs performed by the EPA and ERD.

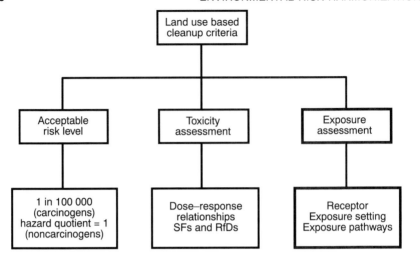

Figure 10.1   Factors affecting cleanup criteria. SFs, slope factors; RfDs, reference
doses.

The development of toxicity values has the least impact on the cleanup
criteria. In general, the values used by both the ERD and the EPA are very
similar as the ERD relies on the EPA Integrated Risk Information System (IRIS)
database and the Health Effects Assessment Summary Tables (HEAST) as its
primary sources for such values. However, some differences remain between
the ERD and EPA in the use of uncertainty factors, and in the number of significant figures used in the toxicity values. These differences are due to policy
decisions made within the ERD to maintain consistency with other aspects of
the cleanup program, such as the reporting of analytical results from environmental laboratories.

The most significant difference between EPA and ERD RAs relates to the
exposure assessment component. The ERD exposure assessment process is
complicated because Michigan exposure assessments must be performed for a
number of different populations and potential land uses. Although the EPA (US
EPA 1991b) has developed standard default exposure assumptions for various
land-use types, it makes slightly different assumptions than the ERD concerning
the frequency and duration of receptor exposures associated with some land-
use categories. For example, the ERD assumes an industrial worker exposure
duration of 21 years instead of 25 years assumed by the EPA. The ERD bases
its value on more recent labor statistical data, and use of the 90th percentile
value rather than the 95th percentile used by the EPA.

Differences in the assumed frequency of receptor exposures between EPA
and ERD result from consideration of the physical characteristics of the site
together with local meteorological factors. Owing to the climatic conditions in

Michigan, exposure to soil contaminants is not expected to occur during the winter months due to snow coverage or frozen ground. As a result, the ERD's exposure frequency default value for industrial land uses is substantially less than the EPA's.

A significant difference in exposure frequency assumptions between the EPA and the ERD also exists for RAs at commercial land uses. This is the result of characterizing the receptor population differently. The ERD has characterized a receptor population that spends less time outdoors than industrial workers. As a result the ERD's exposure frequency values for contacting soil contaminants are considerably lower than the EPA's.

A further disparity in exposure assessment relates to the human exposure pathways considered by the ERD and EPA. The ERD currently considers both ingestion and dermal pathways when developing soil criteria, while the EPA incorporates the ingestion and inhalation routes of exposure. Part 201 generic soil inhalation criteria for ambient air are currently being developed.

The translation of the toxicity and exposure values into land-use-based cleanup criteria is also affected by legislative mandates. For example, the acceptable carcinogenic risk level established in Part 201 is one additional cancer in 100 000 exposed individuals (10^{-5}). This carcinogenic risk level differs from the 10^{-6} risk that the EPA utilizes in the Superfund cleanup program. As one would expect, this translates into an order of magnitude difference in Part 201 criteria compared with Preliminary Remediation Goals (PRGs) developed in the EPA Superfund program.

10.8 IMPACTS ON MEDIA-SPECIFIC CLEANUP CRITERIA

The State of Michigan established land-use-based criteria for chemicals in both groundwater and soil at hazardous waste sites. The determination of the Part 201 drinking water criterion for each chemical is a complex process. If a drinking water standard has been established under the Michigan Safe Drinking Water Act (Act 399, Public Acts of 1976) or an aesthetic value has been promulgated under Part 201, this is used as the criterion value. If neither of these is available, the EPA Superfund RA guidance (US EPA 1989) is used to calculate the drinking water value. However, the application of the guidance has been modified for use in Michigan. The ERD considers exposure only through ingestion while the EPA may require evaluation of potential inhalation exposure pathway risks from household use of the drinking water. In addition, cumulative risks from exposures to multiple contaminants through multiple pathways are not considered by the ERD unless the interaction of the chemicals can be shown to cause greater toxicity than the substances individually.

There are currently two types of soil criteria: one for the protection of groundwater used for drinking water, and another to protect against exposure by direct soil contact. Different approaches are used for developing the criteria and different types of inconsistencies between the EPA and ERD have arisen with respect to each criterion.

To determine soil criteria for the protection of drinking water, there is a default option specified in Part 201 that sets the soil criterion equal to 20 times the drinking water value. However, the ERD is now evaluating the recently released EPA Soil Screening Guidance (US EPA 1996) to pursue development of soil criteria that reflect the fate and transport characteristics of contaminants in soil.

In developing soil criteria for direct contact, the ERD accounts for exposure through both the ingestion and dermal routes. This differs from the EPA approach which combines inhalation and ingestion exposures (US EPA 1991a). However, the recent EPA Soil Screening Guidance provides for the develop-ment of pathway-specific screening levels (US EPA 1996). In the future, the ERD will be assessing if the ingestion and dermal pathways should have separate criteria.

There are specific legislative provisions in Part 201 that apply to a few select chemicals. For example, cleanup of soils contaminated with polychlorinated biphenyls must be performed in compliance with the Toxic Substances Control Act. For chemicals with sufficient data to model their pharmacodynamics, biologically based models may be used to establish cleanup criteria, provided that the EPA has determined the application of the model to be appropriate. Lead is currently the only chemical for which criteria are developed using this approach.

Another area of inconsistency between the ERD and EPA related to deter-mining soil cleanup values is the incorporation of contaminant bioavailability into the generic equations. The ERD applies default absorption efficiency factors to both the ingestion and dermal routes of exposure in calculating soil direct contact criteria, while the EPA considers this factor only for the dermal pathway. This is currently under review by the ERD because toxicity studies rarely attempt to estimate the bioavailability of the administered dose. As a result, it is questionable whether an absorbed contaminant dose should be factored together with an administered dose-based toxicity value in RA calcula-tions.

10.9 CURRENT MICHIGAN ISSUES

The changes in risk management approaches precipitated by the new Part 201 amendments in 1995 resulted in a number of modifications to the development of generic criteria. The majority of changes were targeted to remove perceived excess conservatism from the program and to facilitate consistency among various regulatory standards for contaminants in common media. For example, the assumption that soil contaminants must be remediated to protect ground-water for drinking water use was deleted. Instead, the 'vulnerability' or use of the groundwater aquifer and its potential to support a drinking water supply must be taken into account in establishing a soil criterion cleanup. This change requires the ERD to consider other routes of human exposure to groundwater, such as the dermal or inhalation pathways. As a result, the migration of soil contaminants to groundwater remains a significant health concern. The ERD is currently evaluating approaches to support development of dermal and inhalation exposure-based criteria for cases where the drinking water or ingestion pathway is not applicable. The ERD is using the EPA dermal exposure assessment document and other relevant EPA documents as the bases for developing a Michigan-specific approach.

10.10 CONCLUSIONS

It is evident that a number of differences exist between the ERD and EPA in their approaches to establishing cleanup criteria/goals for hazardous substances at sites of environmental contamination. Some of these differences are simply due to risk management decisions, e.g., the setting of acceptable carcinogenic risk levels. Others result from modifications made to tailor the RA process to reflect the unique conditions present in Michigan, which can affect exposure to environmental contaminants, such as snow cover. This suggests that it is not appropriate to standardize all aspects of the RA process for application nationally. Rather, harmonization may be attained through customizing commonly agreed upon frameworks for RA and risk management to meet the needs of specific states and/or regions.

PART III   A PERSPECTIVE ON RISK MANAGEMENT
(Charles P. Cubbage)

The Michigan Department of Agriculture (MDA) utilizes the risk or tolerances developed by both state and federal agencies. However, the MDA's primary focus is not on risk assessment, but rather, risk management. The author recognizes that risk assessments are commonly used for a variety of purposes, including bureaucratic regulatory needs, public health, and protection of our natural resources. To the extent that risk assessment contributes to reduction of risks either for the public or our natural resources, it offers a valuable tool. However, risk management decisions must often be made in the absence of consistent, or sometimes fairly divergent risk estimates which not surprisingly serve the vested interest of different stakeholders. Under these circumstances, the time and energy spent on resolving the risk assessments differences may be a waste on limited resources.

The MDA has opted to recognize the interests of stakeholders and to address as often as possible risk preventative programs, e.g.: Farm*A*Syst, a voluntary self-assessment of on-farm practices; Clean Sweep, to manage banned and unusable agricultural pesticides; Hazard Analysis of Critical Control Points, that identifies food safety opportunities from the farm to the table; and Integrated Pest Management (IPM) involving economic and pest thresholds.

Historically, the MDA became part of the state effort to coordinate risk assessment, along with the Michigan Department of Natural Resources (MDNR) and the Michigan Department of Public Health (MDPH) during the PBB crisis. The state legislature, dissatisfied with the three agencies' response to the situation created the Michigan Toxic Substance Control Commission with adversarial oversight functions. In 1979 with an Environmental Protection Agency (EPA) grant to the MDNR the three agencies embarked on an effort to produce an interagency risk assessment process. In the mid-1980s the effort was revisited and updated with a final report. In the late 1980s the Michigan Council on Environmental Quality established a risk coordinator position at MDPH under a memorandum of understanding signed by the MDNR, MDPH, and MDA.

In spite of these activities, areas of conflict remain problematic, largely in the arena of interagency inconsistencies. The fish advisories with Food and Drug Administration tolerances versus MDNR derived standards, cleanup of agricultural properties, persistent contamination of compost represent some of the issues at stake. On the other hand, notable risk management success have occurred primarily when the agency extended risk management to include the participation of multiple stakeholders. Although risk management decision authority is legislatively the agency's alone, the responsibility for success can be directly related to the extent to which stakeholders have been made part of the process.

A specific example involving possible misuse of a pesticide on apples illustrates the point. The information on this incident was channeled through a food safety council made up of representatives from over 80 stakeholder groups, including environmentalists, chemical manufacturers, USDA, EPA, MDPH, MDNR, commodity groups, health care providers, and many others. It was a network of recognized leaders from Michigan's diverse communities who explored the many facets of food safety, including their own and others' biases. The group grew into a credible network of two-way communication that was effectively used to share information in response to several potential crises. Within the council, a level of trust had been established to the point where media attention was not seen as a necessary tool to 'get at the truth'. The incident resolution involved cooperative input from multiple state agencies, the US EPA in the form of a special risk assessment, multiple sample analyses from two separate academic institutions, and most importantly, the sharing of the ongoing investigation and risk management options with key members of the food safety council. Those who would have the most at stake, the national apple commodity groups and environmental representatives were kept informed and recognized that neither public safety nor economics had been compromised.

The MDA keenly feels that the investment in preventative risk management and risk communication activities are indispensable tools that can assist policy makers even when risk assessment models leave far more uncertainty and inconsistency than any of us would prefer. At the same time, whenever epidemiological evidence can be gained, it is important to do so. Relevant sample sizes are seldom available in the small populations associated with contamination sites when risk assessment values attempt to protect for the one in a million risk range.

PART I REFERENCES

Agency for Toxic Substances and Disease Registry (1993). *Great Lakes Human Health Effects Research Program: Program Description*. US Department of Health and Human Services, Public Health Service, Washington DC.

Agency for Toxic Substances and Disease Registry (1995). *Effects of Great Lakes Contaminants on Human Health: Report to Congress*. US Department of Health and Human Services, Public Health Service, Washington DC.

Institute for Environmental Studies (1992). *Compendium of Health Effects Databases in the Great Lakes Region*, Vols I and II. Madison, WI.

International Joint Commission (1981). *Proceedings: Workshop on the Compatibility of Great Lakes Basin Cancer Registries*. Health Effects Committee of the Science Advisory Board, IJC, Windsor, Ontario.

Michigan Department of Public Health (1994). *Michigan Behavioral Risk Factor Surveillance System: Assessing Risk Factors at the Regional Level*. Lansing, MI.

PART II REFERENCES

US Environmental Protection Agency (1989). *Risk Assessment Guidance for Superfund*, Vol. 1. *Human Health Evaluation Manual* (Part A). Interim Final. EPA/540/1-89/002. Office of Emergency and Remedial Response, Washington, DC.

US Environmental Protection Agency (1991a). *Risk Assessment Guidance for Superfund*, Vol. 1. Part B. EPA 9285.7-01B. Office of Emergency and Remedial Response, Washington, DC.

US Environmental Protection Agency (1991b). *Human Health Evaluation Manual, Supplemental Guidance: Standard Default Exposure Factors*. OSWER Directive 9285.6-03. Office of Solid Waste and Emergency Response, Washington, DC.

US Environmental Protection Agency (1996). *Soil Screening Guidance: User's Guide*. EPA/540/R-96/018. Office of Emergency and Remedial Response, Washington, DC.

11

A Review of the Consistency, Harmonization, and Scientific Basis of Risk Assessment Activities within the California Environmental Protection Agency

THOMAS A. MCDONALD[1], RICHARD A. BECKER[1], DAVID T.-W. TING[1], LAUREN ZEISE[1], JAMES W. STRATTON[2], ROBERT C. SPEAR[3] AND JAMES N. SEIBER[4]

[1]*California Environmental Protection Agency, Office of Environmental Health Hazard Assessment, Sacramento, CA, USA*

[2]*Department of Preventive Services, California Department of Health Services, Sacramento, CA, USA*

[3]*Department of Environmental Health Sciences, University of California, Berkeley, CA, USA*

[4]*University Center for Environmental Sciences and Engineering, University of Nevada, Reno, NV, USA*

11.1 BACKGROUND

11.1.1 LEGISLATIVE MANDATE FOR REVIEW

Regulatory reform of environmental programs is being addressed at national, state and local levels. In California, the State Legislature as part of its movement towards regulatory reform passed into law Senate Bill 1082. One component of

This chapter is a US Government work and, as such, is in the public domain in the United States of America.

that legislation (California Health and Safety Code, Section 57004) mandated an external scientific peer review of the risk assessment practices of the California Environmental Protection Agency (Cal/EPA). It called for the Director of the Office of Environmental Health Hazard Assessment (OEHHA) within the Cal/EPA to convene an advisory committee consisting of distinguished scientists not employed by the agency.

The law stipulated that the committee conduct a comprehensive review of the policies, methods, and guidelines used by the boards, departments, and offices of the Cal/EPA in performing risk assessments. The law further specified that this review should consider: (1) 'whether or not any changes should be made to ensure that the state's policies, methods, and guidelines for the identification and assessment of chemical toxicity are based upon sound scientific knowledge, methods and practices', and (2) 'the appropriateness of any differences between the policies, methods, and procedures employed by the state and those employed by the National Academy of Sciences, the Environmental Protection Agency and other similar bodies'. The law called on the committee to make recommendations concerning these issues to the Director of the OEHHA and the Secretary of the Cal/EPA.

11.1.2   RISK ASSESSMENT ACTIVITIES AT THE CALIFORNIA ENVIRONMENTAL PROTECTION AGENCY

California has prided itself on being at the forefront of environmental issues and has over the years passed a series of environmental laws. As a result of state and federal mandates, a complex web of federal, state, regional, and local administrative structures has evolved. These federal, state, regional, and local agencies work cooperatively to protect environmental resources and regulate the release of pollutants in California. As part of this cooperation, Cal/EPA and other state agencies, in many cases, administer federal laws or work with the US EPA or other national agencies to carry out federal mandates.

The Cal/EPA, the primary agency overseeing environmental and public health risk assessment activities within California, has developed its own methods, guidelines, policies, regulatory standards, and definitions of hazards, most of which are similar to those used by federal counterparts. There are, of course, some differences; the committee investigated these and specific examples are provided below.

The Cal/EPA was created in 1991 as part of a reorganization of state government by Governor Pete Wilson (Governor's Reorganization Plan 1). This action unified California's state-level environmental protection activities and provided a cabinet-level agency, accountable to the Governor. The reorganization created three new departments: the Department of Pesticide Regulation, formerly part of the California Department of Food and Agriculture; and the Departments of Toxic Substances Control and OEHHA, both formerly part of the California Department of Health Services. In addition to these three departments, the Cal/

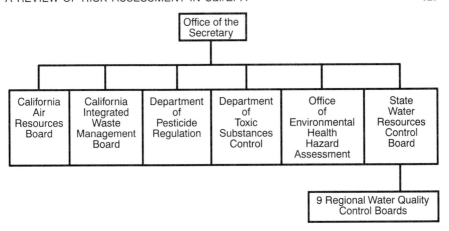

Figure 11.1   California Environmental Protection Agency organizational chart.

EPA includes the California Air Resources Board, the California Integrated Waste Management Board, and the State Water Resources Control Board with its associated nine Regional Water Quality Control Boards (Figure 11.1). OEHHA has the lead responsibility for hazard identification and dose–response assessment of most chemicals, except for pesticides, which are primarily handled by the Department of Pesticide Regulation. OEHHA also has the lead for development of risk guidelines and risk assessments of carcinogens and reproductive toxicants. The California Air Resources Board, the Department of Toxic Substances Control, and the Department of Pesticide Regulation generally conduct exposure assessment, monitoring, fate and transport modeling, and site-specific risk assessment and risk management functions. OEHHA provides toxicological and risk assessment consultation to all Cal/EPA departments as well as other departments in California state government, such as the Department of Health Services and the Office of Emergency Services. A detailed description of the Cal/EPA program activities related to risk assessment can be found in appendix A of the committee's report (Risk Assessment Advisory Committee 1996).

11.2   COMMITTEE MEMBERSHIP AND STRUCTURE

11.2.1   COMMITTEE STRUCTURE

As stipulated in the legislation, the Director of OEHHA was responsible for convening the committee. The primary considerations in establishing the committee were: (1) a manageable time frame for the review period; (2) a wide range of expertise that reflected the multidisciplinary nature of risk assessment

to review the relevant scientific issues; and (3) a size that avoided the ineffi-
ciencies associated with a large committee. A unique structure for the committee
was adopted in order to address the comprehensive nature of the review. A core
group of five scientists was appointed to oversee and guide the entire review.
For the evaluation of specific issues areas, core members were joined by four to
seven experts who were appointed to provide depth of knowledge on a specific
topic. The specific topic areas were: hazard identification; dose–response assessment; exposure assessment; and uncertainty, variability, and risk characterization.
An expert lead member was appointed for each set of expert members. The
expert lead facilitated the discussion at the individual meetings and oversaw the
drafting and editing of the report chapter corresponding to his or her group's
area of expertise. All committee members were invited to participate in all
meetings. Each topic-specific meeting (e.g., dose–response assessment) was
attended by at least nine committee members. At these meetings, findings and
recommendations were drafted by expert members with the core and other
committee members providing insight and commentary.

11.2.2 COMMITTEE MEMBERSHIP

Owing to the multidisciplinary nature of risk assessment, committee members
were drawn from many different disciplines, including toxicology, epidemiology, public health, engineering, medicine, and statistics. Also, in order to
elicit a range of expert opinions on a particular scientific issue, the committee
membership included distinguished scientists from academia, industry, and
national scientific research institutions (see Table 11.1).

11.3 SCOPE OF THE REVIEW

11.3.1 COMMITTEE MEETINGS AND WORKSHOPS

The committee set out and followed an intensive schedule of 10 meetings
which included public forums to discuss a series of topics, namely: (1) hazard
identification; (2) dose–response assessment; (3) exposure assessment
(including human intake and monitoring, and contaminant fate and transport);
(4) uncertainty, variability and risk characterization; and (5) cross-cutting issues.
The meeting on cross-cutting issues focused on five topics: incorporation of
new science into risk assessment, consistency and harmonization, guidelines,
peer review, and resources and organization.

Early in the review process, the committee also held a workshop to review
six case studies in risk assessment (selected from those nominated by the Cal/
EPA and the public). This workshop helped the committee to identify
important issues for further investigation at the topic-specific meetings.
Additionally, the committee held a meeting at the beginning of the process to

Table 11.1 Risk Assessment Advisory Committee

Core members
Herschel Griffin, M.D., California State University San Diego, emeritus
Judith MacGregor, Ph.D., D.A.B.T., consultant and California State University
 San Diego, adjunct
John Moore, D.V.M., D.A.B.T., Institute for Evaluating Health Risks, Washington,
 DC
James Seiber, Ph.D., University of Nevada, Reno and University of California at
 Davis, emeritus (Chair)
Robert Spear, Ph.D., University of California at Berkeley (Vice-Chair)
Experts in hazard identification
Andrew Hendrickx, Ph.D., University of California at Davis
Charles Lapin, Ph.D., D.A.B.T., ARCO
Thomas Mack, M.D., M.P.H., University of Southern California (Expert Lead)
Ronald Melnick, Ph.D., National Institute of Environmental Health Sciences,
 North Carolina
John Peters, M.D., M.P.H., Sc.D., University of Southern California
Richard Thomas, Ph.D., D.A.B.T., International Center for the Environment and
 Health, Virginia
Experts in dose–response assessment
Richard Clark, Ph.D., D.A.B.T., UNOCAL
Kenny Crump, Ph.D., ICF Kaiser, KS Crump Division, Louisiana
Clay Frederick, Ph.D., D.A.B.T., Rohm & Haas, Pennsylvania
Jerold Last, Ph.D., University of California at Davis (Expert Lead)
Christopher Portier, Ph.D., National Institute of Environmental Health Sciences,
 North Carolina
*Experts in exposure assessment 1. Human intake parameters, inter-media
transfers and exposure monitoring*
Hoda Anton-Culver, Ph.D., University of California at Irvine
Gladys Block, Ph.D., University of California at Berkeley
Alison Cullen, Sc.D., University of Washington (Expert Co-lead)
Howard Maibach, M.D., University of California at San Francisco
Fumio Matsumura, Ph.D., University of California at Davis
William Nazaroff, Ph.D., University of California at Berkeley (Expert Co-lead)
Wayne Ott, Ph.D., Stanford University
Paul Price, M.S., ChemRisk, Maine
Experts in exposure assessment 2. Contaminant fate and transport
Steve Brown, Ph.D., Risks of Radiation and Chemical Compounds (R2C2)
Steve Gorelick, Ph.D., Stanford University
Andre Journel, Ph.D., Stanford University
William Nazaroff, Ph.D., University of California at Berkeley (Expert Lead)
Kent Udell, Ph.D., University of California at Berkeley
Akula Venkatram, Ph.D., University of California at Riverside
William Yeh, Ph.D., University of California at Los Angeles
Experts in variability, uncertainty and risk characterization
Thomas Burke, Ph.D., M.P.H., Johns Hopkins University, Maryland
Adam Finkel, Sc.D., U.S. Occupational Safety and Health Administration,
 Washington, DC
William Walker, M.D., Contra Costa County Health Services Department
James Wilson, Ph.D., Monsanto (retired) and Resources for the Future,
 Washington, DC (Expert Lead)

---

plan the scope and format of the review and two working sessions at the end of the review to synthesize its findings and discuss its recommendations for the final report.

All meetings were conducted in accordance with the open-meeting practices of the Bagley-Keane Act (California Government Code Section 11120 *et seq.*). All committee meetings were announced in advance through public notices and were fully open to the public. Public participation was actively encouraged. Committee members, invited panelists, and members of the public actively participated in the discussions at every meeting. Many individuals from the boards, departments and offices of Cal/EPA, federal US EPA, US EPA Region IX, as well as the regional and local air districts that are not formal components of Cal/EPA, also provided important input to the process.

The committee acknowledged its charge to compare Cal/EPA risk assessment practices with those of the National Academy of Sciences (NAS), but noted that the NAS does not routinely practice nor has it adopted procedures for risk assessment. Because the NAS provides information on state-of-the-art risk assessment practice, the committee decided to use the NAS reports as an indication of the state of the art, and in that manner respond to its legal mandate. A report frequently referred to by the committee for this purpose was *Science and Judgment in Risk Assessment* (National Research Council 1994).

11.3.2 LIMITATIONS TO THE SCOPE OF THE REVIEW

The committee focused its review on the present practices of the Cal/EPA boards and departments involved in risk assessment. Although information of a comparative nature was obtained from the US EPA, US Department of Energy, NAS, and from other California State government and regional entities, the committee recognized some limitations to the scope of the review:

'Information on the NAS and the US EPA risk assessment policies and practices was obtained from representatives of the organizations that attended the meetings, policy documents, examples of risk assessments, and the knowledge of the committee members. No in-depth study of these organizations was attempted. The approach of this study emphasized review of the functional components of the risk assessment process and not a detailed department-by-department organizational review.

While several public comments pertained to inconsistencies between regional boards, time did not permit an evaluation of risk assessment issues at the regional level. Issues involving the risk assessment–risk management interface were addressed, whereas those pertaining solely to the risk management process were not evaluated' (Risk Assessment Advisory Committee 1996)

Additionally, the focus of the committee's review was intentionally limited to risk assessment issues related to human health. Ecological risk assessment, which is part of the risk assessment activities of the Cal/EPA, was not addressed, primarily due to ongoing and parallel US EPA and Cal/EPA efforts to develop new ecological risk assessment guidelines which will go through an independent peer review process. However, the committee noted that its findings and recommendations could be applied to ecological risk assessment practices of the Cal/EPA, where appropriate.

11.3.3 DRAFTING OF THE COMMITTEE'S REPORT

The committee directly authored the findings and recommendations of the final report. The primary authors for each topic-specific chapter (e.g., hazard identification, dose–response assessment) were the designated expert members for the meeting covering that topic. One core committee member served as primary reviewer for each topic-specific chapter. The Executive Summary and Introduction were drafted by the core members and discussed at one of the committee's working sessions. The report includes a set of appendices, which were developed by Cal/EPA staff at the request of the committee. A draft report was circulated to the public for review and comment. Written comments from the public on the review draft were considered by the committee in their final revisions to the report.

11.4 COMMITTEE FINDINGS

11.4.1 GENERAL FINDINGS

The committee detailed its findings and recommendations on ways to improve Cal/EPA's risk assessment practices in its final report (Risk Assessment Advisory Committee 1996). In general, the committee found that the risk assessment products of Cal/EPA were of good quality and were credible, both scientifically and professionally. However, the committee noted that there were ample opportunities for improvement and offered a series of recommendations towards that end.

An important theme that the committee repeatedly noted during the course of its review was the inherent conflict between a desire for standardization (e.g., consistency, harmonization, and formalization) of guidelines, policies and procedures and the desire for these same guidelines, policies and procedures to incorporate the latest scientific knowledge and methods. The committee stressed that peer review was a good means to balance these aims.

The committee found many differences in risk assessment practices (e.g., guidelines, policies, methods) both among the various boards and departments within the Cal/EPA and between the Cal/EPA and US EPA programs. Where differences were observed, the primary reasons for the differences stemmed from: (1) differences in state versus federal laws; (2) time of assessment (where one agency had more recent information); or (3) legitimate, defensible differences in scientific judgement in the interpretation of experimental results. However, the reasons for the differences in specific risk assessment products were not as clear.

The committee noted that, in general, a greater degree of similarity existed between like activities of the Cal/EPA and US EPA than existed between different units/departments within the Cal/EPA and US EPA. For example, the Cal/EPA's Department of Pesticide Regulation and US EPA's Office of Pesticide Programs, and Cal/EPA's OEHHA and US EPA's National Center for Environmental Assessment, are more akin in risk assessment methodology than are the Department of Pesticide Regulation and OEHHA. The reasons for this comparability are primarily related to the parallel development of these programs at the federal and state levels. Indeed, formal harmonization of risk assessment practices has been forged between California's Department of Pesticide Regulation and the US EPA's Office of Pesticide Programs through a 'memorandum of understanding'. This agreement facilitates the exchange of pesticide risk assessments and other work products enabling the two agencies to use their resources more efficiently. Recently (as part of the implementation of the committee's recommendations), OEHHA and US EPA's National Center for Environmental Assessment are negotiating a similar 'memorandum of understanding'.

11.5 COMMITTEE RECOMMENDATIONS

11.5.1 GENERAL RECOMMENDATIONS

The committee made 11 general recommendations and over 100 total recommendations to improve the scientific basis and consistency of risk assessment practices at the Cal/EPA. These can be grouped into common themes that are highlighted in the Executive Summary of the committee's report (Risk Assessment Advisory Committee 1996), i.e.: (1) consistency and harmonization; (2) best use of scientific information; (3) the interface between risk assessment and risk management; and (4) organization and management.

11.5.1.1 Consistency and harmonization

The committee recommended that the Cal/EPA take the lead in initiating steps to assure cooperation and consistency with the US EPA. A clearly stated goal for the agency was consistency, where appropriate, in policies, guidelines, methods, and work products, and sharing the workload with the US EPA to maximize resource use. Another recommendation called for the establishment of formal mechanisms, such as an external scientific advisory panel and internal working groups, to promote consistency and harmonization in risk assessment policies and practices, to enhance peer involvement and peer review, and to promote the incorporation of new science into risk assessment practice.

11.5.1.2 Best use of scientific information

The committee recommended that the Cal/EPA provides a forum to identify, evaluate, and promote new or existing scientific knowledge to improve the scientific basis of risk assessment. Related recommendations called for the establishment of *ad hoc* committees to assess specific issues, such as the identification of sensitive populations, receptor-based exposure assessment, indoor air pollutants, and state-of-the-art modeling.

The committee suggested that one of the best ways to promote the use of the 'best' science in regulatory risk assessment was through scientific peer review. The members noted that peer review was unevenly applied in Cal/EPA's technical programs and recommended that the departments of Cal/EPA develop formalized programs for both internal and external peer review. The committee stressed that the level of external peer review should be commensurate with the importance of the risk assessment product or process. For example, routine assessments might not need direct external peer review but the procedures for accomplishing these routine activities should undergo periodic external peer review. Site-specific or chemical-specific assessments, in which a large economic or broad public health impact is anticipated, are good candidates for external peer review.

The committee suggested that the application of new scientific methods and knowledge would be encouraged by continuing education and training of Cal/EPA's technical staff and recommended that Cal/EPA explore ways for scientific staff to participate in continuing education programs and in national and international scientific organizations. Additionally, the committee noted the numerous opportunities within California for increased knowledge through scientific interactions with premier research institutions.

11.5.1.3 The interface between risk assessment and risk management

Historically, the risk decision-making process has been separated into two primary activities: i.e., risk assessment and risk management. These two activities have often been kept separate to provide clear distinction between objective scientific assessment of a risk posed by a hazard (risk assessment) and the social and economic considerations imbedded in risk management. The committee believed that major gains, in terms of program efficiency, could be made by tailoring the level of detail and effort that goes into a risk assessment to the needs of the decision to be made. In the words of a committee member, 'you don't need to goldplate every risk assessment'. This concept is akin to the idea of 'iterative' or 'tiered' risk assessment as described by the NAS committee report *Science and Judgment in Risk Assessment* (NRC 1994) and elsewhere. The committee did not fully embrace the notion of 'iterative' and 'tiered' as these words imply multiple assessments or a reworking of an assessment as one increases in the level of detail (i.e., increases tiers). Instead, the committee suggested a more streamlined approach in which the effort is 'tailored' to or 'balanced' with the importance of the decision.

The committee recommended that the Cal/EPA improve the dialogue between risk assessors and risk managers and other external stakeholders. The committee called on the Cal/EPA to seek input from risk managers and stakeholders, early and throughout the risk assessment process. The committee noted in its discussions that such involvement throughout the process can enhance the overall quality of the assessment by bringing additional concerns, viewpoints, and data to the discussions early in the process, thereby reducing the amount of rework needed later in the process as well as greatly increasing the overall acceptance of the final product.

A recurring theme in many of the committee's discussions was a concern about properly characterizing uncertainty and variability in the risk assessment process in a way that would best help risk managers make informed risk decisions. However, during testimony before the committee, some risk managers indicated a preference for single numbers ('bright lines') on which decisions could be made quickly. The managers noted that the use of ranges in risk estimates, that fall within the bounds of scientific uncertainty, would make their decisions more difficult and could lead to inconsistencies in regulatory actions within or across programs or media. Thus, the committee challenged the agency, through a series of recommendations, 'to better translate emerging methods in risk assessment [such as the evaluation of variability and uncertainty] into risk management policy'. The committee noted that improvements in this area would need to be accomplished through significant interactions between risk assessors, risk managers, and representatives of the public.

11.5.1.4 Organization and management

Regulatory programs to protect the environment, resources, and public health within California have developed over many decades of legislation. A complex web of state, regional, and local structures has developed to meet these legal mandates. The committee felt that it was important that the agency continue to reassess its organizational structure and resources, as well as its legislative underpinnings, to improve and streamline its activities. For example, the committee recommended that the Cal/EPA establish an internal mechanism to ensure that the Secretary of Cal/EPA receives expert advice on a broad range of strategic environmental issues facing Californians. The committee also suggested that the Cal/EPA evaluate its future needs with respect to the scientific disciplines that are required to best meet its mandates and then determine if the staffing and expertise within its programs are appropriate. The committee noted specific staffing needs in fate and transport, exposure assessment, and epidemiology. The committee encouraged the Cal/EPA to develop relationships with other state agencies, the private sector, universities, and other research institutions to meet its needs for specialized expertise not currently available within the Cal/EPA.

11.5.2 SPECIFIC RECOMMENDATIONS

In addition to general themes highlighted in the Executive Summary, the committee provided numerous specific recommendations which were authored by expert committee members. The committee's report includes a complete listing and description of the specific recommendations (Risk Assessment Advisory Committee 1996). The topics of the specific recommendations corresponded to the subject of the committee meetings. They were: (1) hazard identification; (2) dose–response assessment; (3) exposure assessment (including fate and transport modeling); (4) uncertainty, variability risk characterization; and (5) cross-cutting issues. Owing to limitations of space, only a few selected examples will be provided for each topic.

11.5.2.1 Hazard identification

The committee recommended that the Cal/EPA formalize and standardize many of its processes in this area. For example, the Cal/EPA should standardize the collection and submission of pertinent information for hazard identification as well as the content and construction of its hazard identification products. The committee recommended that the Cal/EPA develop a series of uniform processes to: (1) ensure the incorporation of state-of-the-art knowledge (e.g., through peer review); (2) initiate re-review of work products or processes; and (3) develop a formal process to promote internal consistency. The committee made specific recommendations in other areas, including defining adverse effects, incorporating mechanistic data, and developing guidelines on chemical mixtures and sensitive populations.

11.5.2.2 Dose–response assessment

While the committee found the no-observed-adverse-effect level/lowest-observed-adverse-effect level (NOAEL/LOAEL) approach appropriate, they made a series of recommendations regarding the use of uncertainty factors in the development of toxicity values for noncancer endpoints. The committee felt uncomfortable with the standard practice of applying a series of uncertainty factors to poor data, because of the relatively large overall factor that could result. The committee challenged the Cal/EPA to explore alternative ways to bridge gaps in toxicity data, and suggested that overall uncertainty factors greater than 1000 should be justified more extensively. The committee also recommended that the Cal/EPA should develop guidelines on the use of uncertainty factors, which consider the impact of existing data and the severity of the effect. The guidelines should also consider setting a cap on the overall magnitude of uncertainty. If this factor is exceeded, the data base should be considered unreliable or insufficient. Collection of additional data would be indicated.

11.5.2.3 Exposure assessment

The committee recommended that, where appropriate, the Cal/EPA should consider shifting its emphasis from conducting source-based exposure assessments to more receptor-based exposure assessments. Receptor-based exposure assessment methodology involves monitoring exposures to individuals as well as related behavioral patterns, such as time spent indoors versus outdoors. This type of assessment, although resource intensive, provides more reliable data for assessing exposures for individuals. The committee noted that the US EPA has gained some experience in this area resulting from its Total Exposure Assessment Methodology Program. The Cal/EPA has sponsored a smaller number of similar assessments.

11.5.2.4 Uncertainty, variability, and risk characterization

The committee made a series of recommendations calling for the Cal/EPA to review the impacts of its choices of assumptions, models, and use of quantitative uncertainty analysis on the final risk decision. An underlying theme throughout this chapter has been the importance of improved understanding and communication of uncertainty and variability to technical staff, risk managers, and policy makers as well as the public.

1.5.2.5 Cross-cutting issues

The core committee members convened a special meeting to address issues that cut across all areas of risk assessment. The issues selected for discussion included incorporation of new science into risk assessment, consistency and harmonization, guidelines, peer review, and resources and organization. In the area of resources and organization the committee noted that the Cal/EPA generates or maintains a number of databases on measured levels of environmental pollutants in California. The committee observed that the information collected for these databases is probably not used to its full potential. They recommended that the Cal/EPA review the mandates that support these databases and investigate ways to change the collection or reporting of these data so that they are more accessible or usable for risk assessment purposes. For example, slight alterations in the method of sample collection or analysis might provide greater utility of the data for purposes beyond those for which they were originally collected.

In summary, the committee conducted an external, scientific peer review of the methods, policies, and procedures that the Cal/EPA uses to conduct chemical risk assessment. The committee provided a series of findings and recommendations on ways to: (1) improve the scientific basis of risk assessment activities at Cal/EPA, and (2) foster consistency and harmonization among Cal/EPA programs and between Cal/EPA, US EPA, and other similar bodies.

The committee expressed strong endorsement of risk assessment as the primary tool for characterizing, quantifying, and prioritizing risks associated with chemical hazards. Thus, it reaffirmed a process which has been an integral part of Cal/EPA and OEHHA for many years. The report stated that California has been, and continues to be, on the right path in safeguarding the health of the population with respect to environmental hazards. Implementation of the committee's recommendations would further improve the state's capability in this area.

With regard to risk harmonization, the committee noted that regulators must constantly try to balance the desire for standardization and harmonization with the desire for incorporating new scientific methods and information. Scientific peer review is a good means to achieve this balance. Additionally, harmonization of state and federal efforts offers significant potential benefits, such as allowing for more efficient use of resources and providing a consistent approach for the regulated community and the public. However, the committee stressed that harmonization involves a two-way transfer of ideas and information, and that federal risk assessments should not necessarily be considered the 'gold standard'.

The recommendations of the committee will be used by Cal/EPA boards and departments to improve their risk assessment activities now and in the future. Many of the recommendations are applicable to other regulatory agencies and are worthy of consideration by other state and federal bodies in their reform efforts.

REFERENCES

National Research Council (1994). *Science and Judgment in Risk Assessment.* National Academy Press, Washington, DC.
Risk Assessment Advisory Committee (1996). *A Review of the California Environmental Protection Agency's Risk Assessment Practices, Policies, and Guidelines.* Office of Environmental Health Hazard Assessment, California Environmental Protection Agency, Sacramento, CA.

12

Federal–State Toxicology and Risk Analysis Committee: its Role and Accomplishments in Risk Harmonization

KIRPAL S. SIDHU[1], NICHOLAS D. ANASTAS[2], MARIA GOMEZ-TAYLOR[3] AND JENNIFER ORME-ZAVALETA[4]

[1]*Division of Health Risk Assessment, Michigan Department of Community Health, Lansing, MI, USA*

[2]*Office of Research and Standards, Massachusetts Department of Environmental Protection, Boston, MA, USA*

[3]*Office of Water, United States Environmental Protection Agency, Washington, DC, USA*

[4]*National Health and Environmental Effects Research Laboratory, United States Environmental Protection Agency, Research Triangle Park, NC, USA*

12.1 INTRODUCTION

Risk is a measure of the probability and severity of adverse effects (Lowrance 1976). In contrast, safety is defined as the degree to which risks are judged acceptable (Lowrance 1976). According to the National Research Council (NRC), risk analysis consists of the use of factual bases to assess health effects due to the exposure of individuals or populations to hazardous materials and situations (NRC 1983). The process involves the evaluation of information about the hazardous properties of the substances, the extent of human exposure to them, and the characterization of any resulting risk (NRC 1994). Risk analysis is not merely a method of conducting analysis; instead it is a systematic approach to organizing and analyzing scientific knowledge and other information about hazardous activities or substances that might pose risk under

This chapter is a US Government work and, as such, is in the public domain in the United States of America.

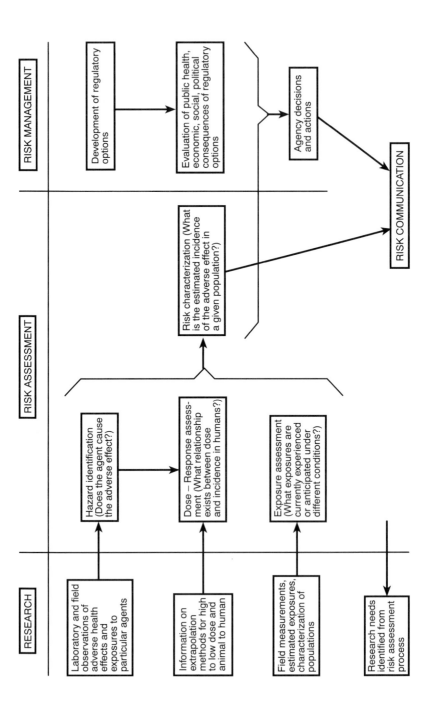

Figure 12.1 Elements of risk analysis. Reprinted with permission from *Risk Assessment in the Federal Government: Managing the Process. Copyright 1983 by the National Academy of Sciences. Courtesy of National Academy Press, Washington, D.C.*

specified conditions (NRC 1994). The elements of risk analysis—risk assessment, risk management, and risk communication—have been described in several publications (NRC 1983, 1994; Covello *et al.* 1988; Scala 1991; Sidhu 1992a; Sidhu and Chadzynski 1994, 1996; Sidhu and Sidhu 1996). The relationships among these elements are illustrated in Figure 12.1.

The NRC has divided the risk assessment process into four steps (NRC 1983). These steps, hazard identification, dose–response assessment, exposure assessment, and risk characterization, have been clearly described in several publications (NRC 1983, 1994; Scala 1991; Sidhu 1992a; Sidhu *et al.* 1994; Sidhu and Chadzynski 1996; Sidhu and Sidhu 1996). Risk assessment procedures are generally used as the scientific bases for setting standards or guidelines for environmental contaminants that may be present in air, food, water, and soil (Cotruvo 1988; Sidhu 1991; Ohanian 1992; Orme-Zavaleta 1992; Sidhu 1992b; Thomas 1992). The application of the risk assessment process to the derivation of standards and guidelines for environmental contaminants requires a number of assumptions and may be accomplished by a variety of methods. As a result of differences in assumptions and methods, there are major differences in the standards and guidelines derived by various regulatory agencies. For example, the drinking water guideline for 1,1-dichloroethane varies from 5 to 850 µ/L (p.p.b.) in different states (Paull *et al.* 1991; Federal–State Toxicology and Risk Analysis Committee (FSTRAC) 1995).

Risk harmonization is a way to address some of the disparities in the risk assessment approaches used by various federal and state regulatory agencies. Risk harmonization is thought of by many as agreement by risk assessors on one particular model and/or set of assumptions. Others have suggested that, while a preferred model might be agreed on for use in risk assessment, risk characterization should incorporate a number of alternative approaches so that several risk estimates can be presented. This will provide the risk manager with important information about the uncertainties of the risk characterization and the range of options available for making a decision (NRC 1994).

12.2 THE FEDERAL–STATE TOXICOLOGY AND RISK ANALYSIS COMMITTEE

The objective of this chapter is to describe briefly the functions of the FSTRAC with a special emphasis on its role and accomplishments in risk harmonization. This committee was organized by the US Environmental Protection Agency (EPA) in 1985 as the Federal–State Toxicology and Regulatory Alliance Committee and was renamed the Federal–State Toxicology and Risk Analysis Committee in 1993. The committee is comprised of representatives of state health and environmental agencies, the Health and Ecological Criteria Division of the EPA, and regional offices of EPA. Representatives from academia and industry have been encouraged to collaborate with FSTRAC by participating in the ongoing projects of the subcommittees, attending meetings, performing document reviews, and conducting surveys (US EPA 1992).

Although FSTRAC has no specific legal mandate, its aim is to encourage interaction and cooperation that can lead to important exchanges of information and experiences in risk analysis, risk harmonization, and regulatory matters. The EPA sponsors this committee for a number of reasons (US EPA 1992).

1. To foster cooperation among states, and between the states and the EPA.
2. To provide a setting for informal and formal discussion of common problems.
3. To improve consistency between federal and state approaches to setting standards or guidelines for environmental contaminants found in drinking water, surface water, and fish tissues.
4. To obtain feedback on proposed federal guidance and standards.
5. To conduct research on factors influencing state risk assessment and regulatory programs.

The first meeting of FSTRAC, held in 1985, concentrated on cooperative efforts among federal and state regulators, toxicologists, and risk assessment specialists to address drinking water issues. However, in recent years the committee has expanded it agenda to include discussions of federal and state toxicology and risk analysis concerns pertaining to the Clean Water Act and fish advisories as well as to the Safe Drinking Water Act (US EPA 1992).

The committee holds its meetings semiannually during fall and spring. Each biannual meeting lasts about two and a half days. The meetings are generally held at EPA headquarters (Washington DC area) or at the EPA regional offices.

At the biannual meetings several scientific presentations are made dealing with new research and ongoing standard-setting activities, and current and proposed regulations are discussed. Each meeting provides an excellent forum to discuss issues, solutions to problems, and specific standard setting approaches developed by the states and the EPA (US EPA 1992). These discussions promote harmonization of risk analyses conducted by federal and state agencies.

FSTRAC has two subcommittees: (1) the Toxicology and Risk Assessment Subcommittee, and (2) the Contaminant Policy and Communication Subcommittee. In addition to meeting at the semiannual gatherings, these subcommittees often hold teleconferences to discuss specific topics. These usually include electronic seminars delivered by national experts.

Working together, the members of FSTRAC have produced several valuable resource materials and documents; those important to risk harmonization are listed here.

1. *Summary of State and Federal Drinking Water Standards and Guidelines, 1993–95* (FSTRAC 1995). This document consists of three separate reports:

Report 1. General information by agency. This provides a description of the drinking water programs in the individual states and the EPA.

Report 2. Detailed standards and guidelines by agency. This lists by agency the current standards and/or guidelines for drinking water.

Report 3. Detailed standards and guidelines by chemical. This lists a number of chemical contaminants, the agencies which have promulgated standards and/or guidelines for each contaminant and the current standard/or guideline levels for each contaminant in drinking water.

2. 'National survey of drinking water standards and guidelines for chemical contaminants' (McGeorge *et al.* 1992). This paper describes the most commonly regulated contaminants and the range of the state and EPA standards or guidelines for drinking water.

3. *Risk Assessment Methodologies: Comparing State and EPA Approaches* (US EPA 1990). This is a report on the Risk Assessment Methodologies Workshop (April 1990) cosponsored by FSTRAC and the EPA. The document compares risk assessment methods and approaches for carcinogens and noncarcinogens used by the EPA and states for drinking water contaminants. Scientists, regulators, and administrators representing different perspectives in federal and state drinking water regulatory agencies participated in this workshop.

4. 'A survey and analysis of states' methodologies for deriving drinking water guidelines for chemical contaminants' (Paull *et al.* 1991). This research paper is an analysis of risk assessment methods used by states to derive drinking water guidelines for carcinogenic and noncarcinogenic environmental contaminants. The data were collected via a survey questionnaire that asked about methods, uncertainty factors, information sources, acceptable risk policies, etc. All the 50 states were analyzed.

Documents 3 and 4 have been particularly useful in risk harmonization efforts (Clarke 1996). In addition to these documents, the state drinking water standards and guidelines have been incorporated into the online computer database, the Hazardous Substances Data Bank (HSDB), a part of the National Library of Medicine's TOXNET system (Fonger 1994). This centralized source also promotes consistency in risk assessment and management across government units.

12.3 RISK HARMONIZATION EFFORTS BY THE FEDERAL–STATE TOXICOLOGY AND RISK ANALYSIS COMMITTEE

At each biannual meeting of FSTRAC, risk assessments for specific chemicals such as dioxin, polychlorinated biphenyls, and methyl tertiary butyl ether, are addressed as well as risk values for such chemicals published in the Integrated Risk Information System (IRIS), Health Effects Assessment Summary Tables

Table 12.1 Risk assessment formulations for environmental contaminants detected in drinking water (FSTRAC 1995)

Contaminant	CAS no.	Agency	Standard/guideline (µg/L
Boron	7440-42-8	ME	600[b]
		MI	600[c]
		MN	600
		EPA	600
Carboxin	5234-68-4	ME	700
		MN	700
		EPA	700
1,4-Dioxane[a]	123-91-1	CT	2
		ME	7
		MI	2
		MN	3
		NC	7
		EPA	3
		EPA	7
Disulfoton	298-04-4	AZ	0.3
		ME	0.3
		MN	0.3
		EPA	0.3
Fluometron	2164-17-2	ME	90
		MN	90
		EPA	90
Tetrahydrofuran	109-99-9	MI	150
		NH	154
		EPA	—

[a]The standard/guideline values have been expressed at a risk level of 1×10^{-6}.
[b]Maine Department of Human Services, Personal Communication.
[c]Sidhu (1995).

(HEAST), Drinking Water Health Advisories, and other related EPA documents (US EPA 1989a,b, 1993, 1994; Orme-Zavaleta *et al.* 1993). In addition, general discussions are held on topics such as carcinogen risk assessment. Discussions on this topic may deal with scaling factors, weight of evidence, guidelines for risk assessment including those for group C carcinogens, mathematical (Linearized Multistage, Logit, Probit, Weibull, etc.), and biological (Physiologically Based Pharmacokinetic) models for extrapolation from high to low dose, and oral and inhalation slope factors. Topics discussed with regard to noncarcinogen risk assessment procedures may include toxicological endpoints, critical effects, no-observed-adverse-effect-levels (NOAEL), lowest-observed-adverse-effect-levels (LOAEL), benchmark doses, uncertainty factors, oral reference doses (RfD), inhalation reference concentrations (RfC), and drinking water specific relative source contributions.

The results of these risk harmonization efforts are reflected in the sample drinking water standards and guidelines derived by several state agencies as shown in Table 12.1. These contaminants were chosen as examples because no federal standards (maximum contaminant levels, MCLs) existed for them, but state standards or guidelines were available. The derivation of the standards/ guidelines for boron, 1,4-dioxane, and tetrahydrofuran (examples 1–3) illustrates the impact of FSTRAC on risk harmonization. The process for boron is provided in detail below to illustrate how the calculations are performed.

12.3.1 EXAMPLE 1: BORON

Health effects of concern: testicular lesions (US EPA 1994)

NOAEL   = 8.8 mg/kg-day (oral diet)

Oral RfD = 8.8 mg/kg-day ÷ 100

$\cong 9 \times 10^{-2}$ mg/kg-day

Interim guideline (Sidhu 1995) = 9×10^{-2} mg/kg-day × 70 kg × 0.20 ÷ 2 L/day

= 0.63 mg/L

= 630 μg/L

= 630 p.p.b.

$\cong$ 600 p.p.b.

where 70 kg represents the default body weight for an adult human, 2 L/day as the default daily water consumption and 0.20 represents the assumed relative source contribution from water. The oral RfD value was derived by the EPA using the NOAEL of 8.8 mg/kg-day from a 2-year study in dogs and applying an uncertainty factor of 100 (US EPA 1994).

Although the drinking water guideline for boron in at least two states (Maine and Michigan) was derived following the EPA approved methodology for noncarcinogens there was some variation in the number calculated using this methodology; it ranged from 600 to 630 p.p.b. However, it has been agreed to round the state of Maine drinking water guideline of 620 p.p.b. to 600 p.p.b. (P. Kemp, Maine Department of Human Services, personal communication, 1996), as Michigan had done, and the two values were then identical.

12.3.2 EXAMPLE 2: 1,4-DIOXANE

The state-derived drinking water values for 1,4-dioxane, based on a risk level of 1×10^{-6} and data from the National Cancer Institute study (NCI 1978), vary from 2 to 7 p.p.b. in different states (Table 12.1). This variation is due to differences in expert judgement which resulted in the use of different sets of data (Sidhu and Sidhu 1996) from the same study (NCI 1978).

12.3.3 EXAMPLE 3: TETRAHYDROFURAN

The interim oral RfD for tetrahydrofuran of 0.022 mg/kg-day was derived by Hurst (1990) and obtained through personal communication by FSTRAC members (Sidhu 1992b). It serves as the basis for drinking water guidelines for the states. The small difference between Michigan and New Hampshire's guideline for tetrahydrofuran (Table 12.1) again represents different decisions as to the appropriateness of rounding off the final value.

12.3.4   EXAMPLE 4: DRINKING WATER HEALTH ADVISORY PROGRAM OF THE ENVIRONMENTAL PROTECTION AGENCY

Another process that promotes consistency among the state and federal drinking water standards and guidelines is the EPA's Drinking Water Health Advisory Program (Orem-Zavaleta *et al.* 1993). Through this program, the EPA develops technical guidance on the short-term and long-term health effects of contaminants in drinking water. FSTRAC serves as a vehicle for the review of draft health advisory documents that are of interest to specific states, particularly when short-term exposures can occur from a spill, or for contaminants that are found in drinking water but are not regulated. This active participation by FSTRAC members contributes to consistency in the derivation of standards and guidelines by the states.

12.4   DISCUSSION

The FSTRAC biannual meetings and teleconferences have brought officials from the states and the EPA together to discuss mutual concerns and to establish a network of national communications (US EPA 1992). This network facilitates risk harmonization among various state agencies involved in risk assessment and risk management by enhancing personal communication among state and federal officials directly involved in these activities. These communications address both scientific knowledge in toxicology and risk analysis procedures.

    The discussions held by FSTRAC members on: assumptions and methods for conducting risk assessment for noncarcinogens; FSTRAC publications and compilations of resource materials; utilization of EPA derived and officially approved data on slope factors, RfDs, and RfCs contained in HEAST and IRIS (US EPA 1993, 1994); and the application of EPA recommended assumptions, uncertainty factors, and relative source contribution factors for drinking water all contribute to risk harmonization. The participation of FSTRAC members in the review of drafts of drinking water health advisories (Orme-Zavaleta *et al.* 1993) also promotes consistency in the derivation of standards and guidelines by states.

    Although FSTRAC has facilitated harmonization (Table 12.1), there remain some differences in risk assessment and risk management procedures between individual states and between the states and the EPA due to differences in assumptions and approaches. For example, while most of the states manage cancer risk at an acceptable risk level of 1 (additional cancer) in 1 000 000 exposed individuals (1×10^{-6}), a few states manage cancer risk at an acceptable risk level of 1×10^{-5}. For another, states may include different routes of exposure in their assessments. In addition, state assessors may utilize a variety of assumptions and approaches to provide risk managers with information about assessment uncertainties and knowledge of the range of options available in decision making. Further, some assessments may be site-specific. For example, in the implementation of Great Lakes fish advisories (US EPA 1995), states not only use different cancer risk levels but also use fish consumption rates that may vary from one location to another. Therefore, even if the methodologies are consistent, the resulting numerical values may not always be the same.

In addition, there are rare examples where states have modified the EPA-derived RfD values (US EPA 1993) for use in their risk management programs. For example, one state modified the EPA oral RfD value of carbon disulfide from 0.11 to 0.011 mg/kg-day to derive a drinking water guideline for this contaminant (Sidhu 1992b). An additional uncertainty factor of 10 was applied and justified by the state because carbon disulfide is cardiotoxic in addition to the other health effects considered by the EPA, such as neurotoxicity and developmental toxicity (Sidhu 1992b; US EPA 1993; Agency for Toxic Substances and Disease Registry (ATSDR) 1995). It is interesting to note that the ATSDR has set its minimal risk level (MRL) for carbon disulfide at 0.01 mg/kg-day (ATSDR 1995), which is consistent with the state oral RfD value (0.011 mg/kg for day) for this contaminant (Sidhu 1992b).

When there is no federal health advisory level or there are no suitable data available for conducting a risk assessment for a drinking water contaminant found in a state, the state toxicologist and risk assessment specialist may be required to set an interim guideline for this contaminant within a short period of time, i.e., 7–10 days, to decide if remedial action is necessary to protect human health. As the EPA usually requires several years to set a drinking water standard, the state toxicology and risk assessment specialist has to use whatever data are available and exercise professional judgement in issuing an interim guideline for the drinking water contaminant. In such cases there may be disparities among states in risk assessment and risk management.

Participation in biannual meetings by state representatives requires funds and time. Lack of travel funds and restrictions on out-of-state travel sometimes contribute to reduced participation by state representatives in the FSTRAC meetings. This is unfortunate because FSTRAC has contributed toward solving toxicology and risk analysis problems in the implementation of the Safe Drinking Water Act, and has significant potential to initiate similar activities for issues involved in the implementation of the Clean Water Act. In addition, FSTRAC is a two-way street for communication between state and federal agencies and facilitates state input on federal policy. Participation of state as well as federal representatives is vital for achieving the goals of FSTRAC. Therefore, there is a need to address seriously the problem of lack of participation of state representatives in FSTRAC meetings. One important outcome of increased participation of states in FSTRAC meetings and projects will be enhanced risk harmonization.

12.5 CONCLUSIONS

The FSTRAC has contributed greatly toward risk harmonization especially in the area of drinking water standards and guidelines. Some of the accomplishments (US EPA 1990; Paull *et al.* 1991) of the committee have been nationally recognized (Clarke 1996). The committee may be considered as a role model for interactions among other federal and state agencies involved in risk analysis.

ACKNOWLEDGMENTS

The senior author expresses his sincere thanks to Professor Michael A. Kamrin (Institute of Environmental Toxicology, Michigan State University, East Lansing) and the Organizing Committee of the Conference on Harmonization of State/Federal Approaches to Environmental risk (May 20–21, 1996), for their kind invitation to participate in the conference and to contribute this invitational paper to the proceedings. We also thank Daniel A. Dolanski, D.O., James W. Bedford, Ph.D., Peter T. Kliejunas (State of Michigan) and Scott Stoner (State of New York) for their review of the manuscript.

REFERENCES

Agency for Toxic Substances and Disease Registry (1995). *Toxicological Profile for Carbon Disulfide, Update.* US Department of Health and Human Services, Public Health Service, Washington, DC.
Clarke DP (1996). Analysis: Experts eye risk assessment disparities as states take center stage. Risk Policy Report, April 19, pp. 29–30.
Contaminant Policy and Communication Subcommittee, Federal–State Toxicology and Risk Analysis Committee (FSTRAC) (1995). Summary of State and Federal Drinking Water Standards and Guidelines (1993–1995). Office of Science and Technology, Office of Water, US Environmental Protection Agency, Washington, DC.
Cotruvo JA (1988). Drinking water standards and risk assessment. *Regulatory Toxicology Pharmacology* **8**: 288–299.
Covello VT, Sandman PM, Slovic P (1988). *Risk Communication, Risk Statistics, and Risk Comparisons: A Manual for Plant Managers.* Chemical Manufacturers Association, Washington, DC.
Fonger G (1994). National Library of Medicine demonstration of online database: The incorporation of federal and state drinking water standards and guidelines into NLM's Hazardous Substances Data Bank (HSDB). Minutes of the FSTRAC Biannual Meeting, November 16–18, 1994, pp 34–35, Washington, DC.

Hurst P (1990). Provisional RfD for tetrahydrofuran (THF). Memorandum, May 3, 1990. US Environmental Protection Agency, Cincinnati, OH.

Lowrance WW (1976). *Of Acceptable Risk: Science and the Determination of Safety*. William Kaufmann, Inc., Los Altos, CA.

McGeorge LJ, Kreitzman SJ, Dupuy CJ, Mintz B (1992). National survey of drinking water standards and guidelines for chemical contaminants. *Journal of the American Water Works Association* **84**:72–76.

National Cancer Institute (1978). *Bioassay for 1,4-Dioxane for Possible Carcinogenicity*. DHEW Pub No (NIH) 78-1330. National Institutes of Health, US Department of Health, Education and Welfare, Washington, DC.

National Research Council (1983). *Risk Assessment in the Federal Government: Managing the Process*. National Academy Press, Washington, DC.

National Research Council (1994). *Science and Judgment in Risk Assessment*. National Academy Press, Washington, DC.

Ohanian EV (1992). New approaches in setting drinking water standards. *Journal of the American College of Toxicology* **11**: 321–324.

Orme-Zavaleta J (1992). Toxicological basis for drinking water: Unreasonable risk to health values. *Journal of the American College of Toxicology* **11**: 325–329.

Orme-Zavaleta J, Cantilli R, Ohanian EV (1993). Drinking water health advisory program. *Annali dell Instituto Superior di Sanita* **29**: 355–358.

Paull JM, Joellenbeck LM, Cochran RC, Sidhu KS (1991). A survey and analysis of states' methodologies for deriving drinking water guidelines for chemical contaminants. *Regulatory Toxicology and Pharmacology* **13**: 18–35.

Scala RA (1991). Risk assessment. In: Toxicology. The Basic Science of Poisons (Amdur MO, Doull J, Klaassen CD, eds). Pergamon Press, New York, NY, pp. 985–996.

Sidhu KS (1991). Standard setting processes and regulations for environmental contaminants in drinking water: State versus federal needs and viewpoints. *Regulatory Toxicology and Pharmacology* **13**: 293–308.

Sidhu KS (1992a). Current methods of assessment of exposure to environmental contaminants. *International Journal of Toxicology, Occupational and Environmental Health* **1**: 84–96.

Sidhu KS (1992b). Regulation of environmental contaminants in drinking water: State methods and problems. *Journal of the American College of Toxicology* **11**: 331–340.

Sidhu KS (1995). Formulation of risk assessment for deriving a drinking water guideline for boron. Unpublished (May 24, 1995), Michigan Department of Public Health, Lansing, MI.

Sidhu KS and Chadzynski L (1994). Risk communication: Health risks associated with environmentally contaminated private wells versus chloroform in a public water supply. *Journal of Environmental Health* **56**: 13–16.

Sidhu KS and Chadzynski L (1996). Communication of health risks associated with polychlorinated biphenyls contaminated soil near an elementary school. *Technology: Journal of the Franklin Institute* **333A**: 7–15.

Sidhu KS and Sidhu JS (1996). Risk assessment for environmental carcinogens. In: Current Concepts in Human Genetics (Singh JR, ed.). Guru Nanak Dev University Publishers, Amritsar, India, pp. 138–151.

Thomas RD (1992). Toxicologic risk assessment issues for drinking water safety. *Journal of the American College of Toxicology* **11**: 311–319.

US Environmental Protection Agency (1989a). *Risk Assessment Guidance for Superfund*, Vol. 1, *Human Health Evaluation Manual* (Part A). Interim Final EPA/540/1-89/002. Office of Emergency and Remedial Response, Washington, DC.

US Environmental Protection Agency (1989b). *Guidance Manual for Assessing Human*

Health Risks from Chemically Contaminated Fish and Shellfish. EPA/503/8-89/002. Office of Marine and Estuarine Protection, Washington, DC.

US Environmental Protection Agency (1990). *Risk Assessment Methodologies: Comparing State and EPA Approaches.* EPA 570/09-90-012. Office of Drinking Water, Washington, DC.

US Environmental Protection Agency (1992). *EPA FSTRAC: A Network of Knowledge. Federal–State Toxicology and Regulatory Alliance Committee.* Office of Water, Washington, DC.

US Environmental Protection Agency (1993). Integrated Risk Information System: Announcement of availability of background paper. *Federal Register* **58**: 11490–11495.

US Environmental Protection Agency (1994). *Health Effects Assessment Summary Tables, FY-1994 Annual.* EPA 540/R-94/020. Office of Solid Waste and Emergency Response, Washington, DC.

US Environmental Protection Agency (1995). Final water quality guidance for the Great Lakes System: Final rule. *Federal Register* **60**: 15366–15425.

The senior author and co-authors are all members of the FSTRAC, however, the opinions expressed in this chapter are solely those of the authors. These do not necessarily reflect the official policy of the authors' organizational affiliations or any other agency.

13

Case Study: Sport Fish Consumption Advisories in the Great Lakes Region

JOHN L. HESSE

Division of Health Risk Assessment, Michigan Department of Community Health,
Lansing, MI, USA

13.1 BACKGROUND

Sport fish consumption advisories in the Great Lakes region serve as an excellent case study highlighting the need for risk harmonization and the difficulties in attaining agreement when multiple state and federal agencies are involved. These advisories provide sport fish consumption recommendations to anglers so they can limit consumption of fish which are likely to contain unacceptable levels of contaminants. Owing to the location of the Great Lakes on the boundary between the United States and Canada, consistency in advice is an international issue involving both US federal and state agencies as well as Canadian provincial and federal governments.

Ironically, fish consumption advisories have become highly controversial at a time when contaminant levels are at their lowest point in the last 25 years. Since late January 1996, the Michigan Department of Community Health (MDCH) and the Governor of Michigan have been under heavy criticism by various organizations and agencies because of Michigan's recent decisions to relax salmon consumption advisories that have been in place for over 20 years.

After delaying advisory relaxations for several years while awaiting adoption of a uniform protocol among Great Lakes jurisdictions, Michigan made a policy decision in 1995 to once again apply the criteria adopted by the Lake Michigan states in 1985 and which have been in effect since then. Because none of the coho salmon sampled from Michigan waters of Lake Michigan since 1986 had exceeded those criteria and neither had any chinook salmon sampled since 1988, Michigan completely removed advisories for these two species in 1996.

This chapter is a US Government work and, as such, is in the public domain in the United States of America.

The US Environmental Protection Agency (EPA) and some environmental groups criticized Michigan's action as not being protective of public health. In addition to adverse media coverage, Michigan's Governor received a letter from the regional administrator of the EPA petitioning him to retract the relaxation because it represented 'a serious health threat to future generations of people in Michigan'. The Governor also received a letter from the International Joint Commission (IJC), a copy of which had been sent to the US Secretary of State, requesting a reversal of this policy decision. The IJC is an oversight body created by the United States and Canada for protection of the Great Lakes. Quite remarkably, all of this reaction occurred in a situation where not even 1% of fish samples analyzed over the last 8–10 years exceeded the concentration which the US Food and Drug Administration (FDA) currently allows in fish sold in the marketplace.

Such divergent opinions reflect the political reality that various agencies/organizations operate under different missions and mandates to accomplish different objectives. For example, the EPA is a regulatory agency with responsibility for 12 major laws which tries to establish and maintain regulatory consistency across all programs. Though fish consumption advisories are not regulatory actions and do not fall under the authority of the EPA, the agency pursues consistency in advisories because it contributes to consistency in the implementation of their other programs, e.g., the Superfund.

Environmental groups, such as the National Wildlife Federation, consider fish advisories important because they provide an opportunity to pressure government and industry for further reductions in the discharge of toxic chemicals into the environment. They perceive that relaxation of fish consumption advisories based on declining fish contaminant levels could undercut efforts to increase, or even maintain, strong regulatory control over waste discharges and the cleanup of contaminated sites.

Another important consideration is that fish consumption advisories have been the driving force behind many environmental decisions at the state, federal and international levels. Fish consumption advisories are often used to demonstrate 'impaired natural resource use' resulting from a site of contamination. In fact, many current Superfund sites would not have scored high enough to qualify for the National Priority List (NPL) if fish advisories were not in place. Also, many of the 42 IJC Great Lakes Areas of Concern (AOC) would probably not have been designated as AOCs without fish consumption advisories being an issue.

13.2 CHEMICALS OF CONCERN

In 1970, Michigan became the first state to issue fish consumption advisories, following the discovery of mercury in Lake St. Clair fish downstream from a chlor-alkali facility. However, since then advisories have become commonplace and 47 states currently have fish consumption advisories in place.

Table 13.1   Contaminants resulting in fish and shellfish advisories

Contaminant	Number of states issuing advisories
Metals	
Arsenic (total)	1
Cadmium	1
Chromium	1
Copper	1
Lead	5
Mercury	33
Selenium	5
Zinc	1
Organometallics	1
Unidentified metals	4
Pesticides	
Chlordane	23
DDT and metabolites	10
Dieldrin	4
Heptachlor epoxide	1
Hexachlorobenzene	2
Kepone	1
Mirex	3
Photomirex	1
Toxaphene	2
Unidentified pesticides	5
Polycyclic aromatic hydrocarbons (PAHs)	3
Polychlorinated biphenyls (PCBs)	36
Dioxins/furans	22
Other chlorinated organic chemicals	
Dichlorobenzene	1
Hexachlorobutadiene	1
Pentachlorobenzene	1
Pentachlorophenol	1
Tetrachlorobenzene	2
Tetrachloroethane	1
Others	
Creosote	2
Gasoline	1
Phthalate esters	1
Polybrominated biphenyls (PBBs)	1
Fecal coliform bacteria	1
Tributyl tin	1
Multiple pollutants	2
Priority contaminants	1
Organic chemicals	1
Volatile organic compounds	1
Unspecified pollutants	3

US EPA (1995).

153Table 13.1 presents a list of chemicals that were the basis of fish consumption advisories in one or more states as of 1994 (US EPA 1995). Mercury is very commonly the subject of advisories, with 33 states having advisories due to mercury contamination. Advisories based on chlordane (23 states) and dioxins/furans (22 states) are also fairly common. Of most frequent concern, however, are the polychlorinated biphenyls (PCBs), which led 36 states to issue fish advisories in 1994 to manage risk from that class of chemicals. PCBs are the primary contaminant responsible for advisories in Great Lakes' waters as well, and have been the focus of consumption advisory harmonization efforts for several years.

What is known about PCBs in fish and human health? Monitoring of PCBs in human sera in 1974 and 1980 by the Michigan Department of Public Health (MDPH) showed a direct relationship between the amount of fish eaten in pounds per year and the level of total PCBs in the blood (MDPH 1983). Other studies showed that fat and blood levels were related to both the amount of fish consumed per year and the number of years that the contaminated fish were eaten.

Much of the concern regarding health effects of PCBs is focused upon their potential for adverse impacts upon growth and development of young children exposed *in utero* (Jacobson *et al.* 1983; Fein *et al.* 1984; Humphrey 1991). Although some of the studies on PCBs are controversial (Paneth 1991), public health agencies generally believe that protection against possible neurodevelopmental impacts from PCBs on children is a priority.

13.3 CONTAMINANT TRENDS

One important factor contributing to risk management decisions regarding fish consumption is the trend of contaminant levels in fish. Are levels increasing or decreasing, and at what rate? The Lake St. Clair mercury problem serves as a good illustration of the changes in contaminant levels that have been occurring in the Great Lakes basin. The average mercury levels in fish were initially very high, with Lake St. Clair walleye averaging more than 2 parts per million (p.p.m.). As discharges into the lake were halted, levels have decreased to approximately 0.5 p.p.m.; a decline of nearly 80% (Vaillencourt A, personal communication, 1996). Similar declines have been seen for DDT levels in Lake Michigan lake trout, with a drop of about 90% since 1970 (Devault D, personal communication, 1992). DDT was banned in 1969 in Michigan and Wisconsin, and nationally in 1972.

Although PCB levels in fish increased in the early 1970s, a rapid decline occurred after 1975 (Figure 13.1), when Michigan took the lead in banning the use of PCBs. Several other states soon followed and the EPA then took action under the Toxic Substances Control Act (TSCA) banning PCBs from commerce effective on July 1, 1979. PCB levels in Great Lakes' fish have declined by approximately 90% in the past 20 years.

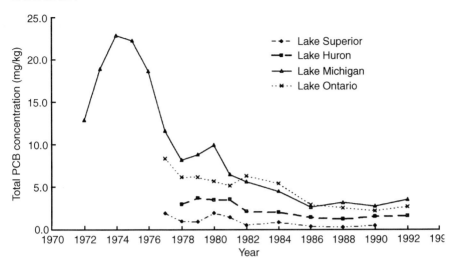

Figure 13.1   Mean total polychlorinated biphenyl (PCB) concentrations in whole lake trout from the Great Lakes, 1972–1992 (DeVault *et al.* 1996, in press).

Downward trends in fish contaminant levels are being paralleled by corresponding, but less dramatic, declines of contaminant concentrations in human blood of Great Lakes fish eaters. Hovinga *et al.* (1992) compared levels of PCBs and DDT in blood samples collected from Michigan's Great Lakes fisheater study cohort in 1982 and 1989. During this time, there was a 39% decrease in DDT residues and a small decline in PCB levels. The researchers hypothesize that the reductions in DDT and PCB levels resulted from lower levels in the environment rather than changes in fish consumption patterns.

Figure 13.2 shows what has happened to the author's PCB blood levels in the past 10–15 years. In 1983, the author's average level of 38 parts per billion (p.p.b.) was about eight times higher than that of the general non-fish eating population in Michigan. Through a combination of contaminant level decreases in fish and the author's adherence to consumption advisories, a 50% decrease in this level occurred by 1989. The level further decreased to 12 p.p.b. in 1993 and 8 p.p.b. in 1995. This is very close to the background level in the average individual who doesn't eat sport caught fish, and was achieved while the author consumed frequent meals of sport fish from the Great Lakes and inland waters. This is further evidence that contaminant levels are declining in the environment and that adhering to advisory recommendations can be an effective way of reducing one's exposure to contaminants.

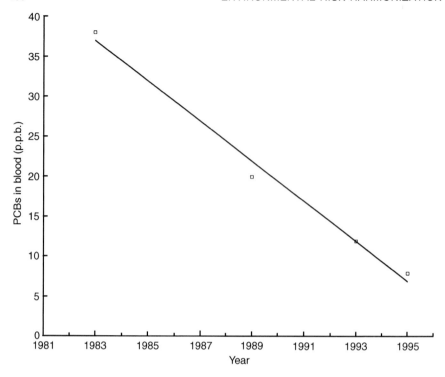

Figure 13.2 Polychlorinated biphenyl (PCB) blood levels of the author, 1983–1995.

13.4 ADVISORY TRIGGER LEVELS

Historically, Michigan has based its trigger levels for fish consumption advisories (Table 13.2) on the FDA action or tolerance levels used to regulate fish sold in the marketplace. However, for two contaminants, Michigan has differed from the FDA. Michigan set the trigger level for dioxin at 10 parts per trillion (p.p.t.) compared with the federal government's unofficial 25 p.p.t. level of concern and decided to use a trigger level of 0.5 p.p.m. for mercury in contrast to the FDA tolerance level of 1.0 p.p.m.

With regard to Michigan's decision to relax the chinook salmon advisory in 1996, how do recent PCB levels in fish compare with FDA acceptable limits? Figure 13.3 displays all of the PCB data for chinook salmon taken from Michigan waters of Lake Michigan since 1990. The salmon tested ranged from 20 inches to about 40 inches in length, a size representative of salmon caught by anglers. These data show an average PCB concentration of about

Table 13.2 List of contaminants and trigger levels currently used by the Michigan Department of Community Health (MDCH) in the establishment of fish consumption advisories

Chemical	MDCH advisory trigger
Chlordane	0.3 p.p.m.[a]
DDT	5.0 p.p.m.[b]
Dieldrin	0.3 p.p.m.
Dioxin	10.0 p.p.t.[c]
Endrin	0.3 p.p.m.
Heptachlor	0.3 p.p.m.
Mercury	0.5 p.p.m.
Mirex	0.1 p.p.m.
Polychlorinated biphenyls	2.0 p.p.m.[d]
Toxaphene	5.0 p.p.m.

[a]Total chlordane isomers and related compounds.
[b]Total DDT and metabolites (DDE and DDD).
[c]Total chlorinated dioxins and furans as toxic equivalents of 2,3,7,8-tetrachlorodibenzodioxin (2,3,7,8-TCDD).
[d]Total polychlorinated biphenyls.
p.p.m. = parts per million; p.p.b. = parts per billion; p.p.t. = parts per trillion.

1 p.p.m. No fish contain PCB levels near the FDA's 2.0 p.p.m. trigger level. In fact, no samples of chinook salmon from Michigan waters have exceeded 2.0 p.p.m. since 1988. For coho salmon (the smaller salmon species in Lake Michigan), this trigger level hasn't been exceeded since 1985. Based on these data, and using the criteria agreed to with the other Lake Michigan states, Michigan decided in 1995 to remove the consumption advisory for both chinook and coho salmon, even though the Great Lakes states were considering adoption of criteria that could be more restrictive.

13.5 UNIFORM ADVISORY CRITERIA

Although all of the Great Lakes states have issued fish consumption advisories for a number of years, the advice has differed from state to state. Recognizing this problem, a number of efforts have been made to harmonize fish consumption advice. In 1983, the four states bordering Lake Michigan (Illinois, Indiana, Michigan, and Wisconsin) joined together to work toward uniformity, a process facilitated by EPA Region V. By early 1985 these states announced the first uniform health advisory on sport fish consumption for

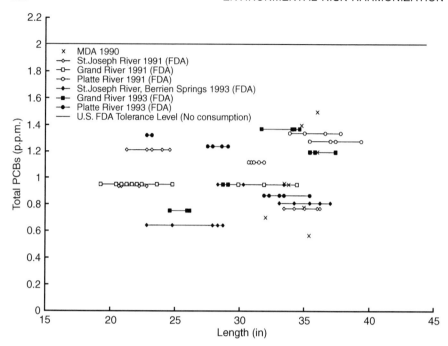

Figure 13.3 Polychlorinated biphenyl (PCB) concentrations versus fish length in chinook salmon from Michigan waters of Lake Michigan, 1990–1993. NB: symbols connected by horizontal lines represent composite samples.

Lake Michigan anglers. In addition to agreeing on maximum acceptable levels for contaminants and advisories for particular fish species, the participants agreed to increase data sharing and coordination of monitoring plans.

These agreements reflected the states' belief that sport anglers required more protection than consumers of fish sold in the market because of their greater consumption of fish. However, they also reflected a reluctance to break completely from the FDA PCB tolerance level of 2.0 p.p.m. The states believed it would be difficult for the public to accept an approach that treats anglers too much differently than people who buy their fish in restaurants and markets. Thus, rather than simply taking the FDA approach and advising no consumption if the average PCB level was over 2.0 p.p.m. and unlimited consumption if the average level was below 2.0 p.p.m., the Lake Michigan states decided to add a 'restrict consumption' category. Fish would be assigned to this category when 11–49% of the samples of that species contained more than 2.0 p.p.m. It was also decided that when 50% or more of the samples exceed 2.0 p.p.m., 'no consumption' would be recommended. This 'no consumption advice' would then be consistent with the FDA regulation of fish sold in the market.

Recognizing the increased sensitivity of some segments of the population to the effects of PCBs, women of child-bearing age and children under 15 were advised not to eat any fish falling into the 'restrict consumption' category. The general population was advised to eat no more than one meal per week of fish in this category.

Figure 13.4 shows how the Lake Michigan states' approach was implemented for chinook salmon in 1987. The figure shows the pooled fish monitoring data from Wisconsin and Michigan plotted according to length. Visual inspection was used to select natural break points at which more than 10% of the samples were over 2.0 p.p.m. and another at which more than 50% of the samples were above this tolerance level. This very simple approach, which is easily implemented and easily understood, has been the basis for the Michigan advisories since 1985.

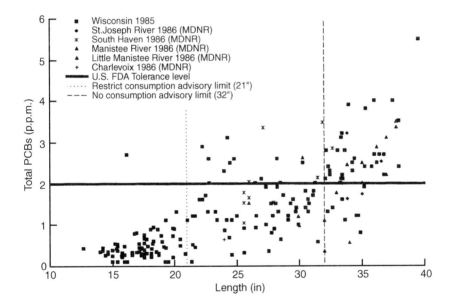

Figure 13.4 Polychlorinated biphenyl (PCB) concentrations versus fish length in chinook salmon from Michigan and Wisconsin waters of Lake Michigan, 1985–1986. Note: The vertical dotted line (..........) represents the division between fish in the unrestricted consumption (less than 21 inches) and restricted consumption categories (between 21 and 32 inches). The vertical dashed line (– – – –) represents the division between the restricted consumption (between 21 and 32 inches) and no consumption categories (32 inches or longer).

13.6   GREAT LAKES TOXIC SUBSTANCE AGREEMENT

Building on actions taken to harmonize Lake Michigan advisories, the Great Lakes governors met in 1986 and adopted the Governors' Great Lakes Toxic Substance Agreement. One of the provisions of this agreement was that all Great Lakes states must reach agreement by 1987 on common fish consumption advisories for all of the Great Lakes. This mandate was interpreted to apply on a lake by lake basis except where regional differences within the lake warranted different advice. This was a difficult task and the states were given a very short time to accomplish it.

13.7   CREATION OF A TASK FORCE

Following the adoption of the agreement, a Great Lakes Fish Consumption Advisory Task Force (hereafter referred to as 'the Task Force') was convened. It consisted of up to three representatives from each state (e.g., one each from the agriculture, health, and environmental agencies) as well as representatives from the FDA, EPA, and the Canadian Directorate of Health and Welfare. In addition, one tribal representative, a representative of the Province of Ontario, and an environmental organization representative participated in the meetings. In total, 27 different organizations from nine jurisdictions, including Canada, were represented in the Task Force.

Surprisingly, the Task Force did satisfy its mandate of developing consistent advisories on each of the Great Lakes by 1987, despite variations in advisory criteria among the states. This was accomplished by looking for commonalities in advice and negotiating agreement where there were differences. Although the Task Force was successful for the 1987 fishing season, it did not reach agreement on the criteria for establishing future advisories. Thus, further meetings were needed to address that issue.

As part of these deliberations, the Task Force identified factors that were critical to achieving harmony or commonality among all of the organizations represented. Table 13.3 lists these factors in the order they were raised by Task Force members during a work session in 1986. Foremost were scientific issues, such as scientific defensibility and adequate protection of public health. However, the Task Force also identified social, legal, administrative, and political issues such as acceptance by the general public, implications for environmental control programs, and impacts on legal mandates. Acceptance by the public was considered critical because of the likelihood of poor compliance with advisories that were confusing or contradictory.

The Task Force also discussed the possibility that significant health benefits of fish consumption could be lost if the advisories led people to stop consuming fish. The group agreed that the primary goal of the advisories was not to discourage fish consumption but to help anglers make better choices

Table 13.3 Critical factors in selection of a fish consumption advisory methodology[a]

Adequate protection of public health (ethical, moral, and legal implications)
Comprehension and acceptance by the general public
Implications for environmental control programs
Implications for the sport and commercial fishing industry
Multimedia exposure to contaminants involved
Sensitive population groups, as well as the average fish consumer
Need for improved consumption rate information for better definition of exposure potential
Opportunity for public review and input
Evaluation of program effectiveness
Historical perspective and contaminant trends
Legal mandates, state rights, federal preemption authority
Definition of 'acceptable' risk
Health benefits of fish consumption
Publicize 'clean' fish as well as ones to avoid
Avoid scare tactics
Availability of adequate funding for fish monitoring and advisory program in all jurisdictions

---

[a]Work Session of Great Lakes Fish Consumption Advisory Task Force (1986).

about which fish they eat and to encourage them to prepare the fish in ways that minimize exposure to contaminants.

13.8   SUMMARY OF EXISTING PROTOCOLS

Early in its deliberations, the Task Force surveyed the approaches of all of the jurisdictions represented. A summary and analysis of existing programs (Hesse 1990) helped the Task Force recognize commonalities (Table 13.4) and focus on differences (Table 13.5) which still needed to be resolved. An example of a unique approach was the decision by New York to consider multiple contaminants when placing fish in consumption categories. To combine these multiple contaminants, New York calculated the ratio of the concentration for each contaminant to the tolerance or action level for that contaminant, and then added the resulting ratios together for all contaminants present.

Some differences that the Task Force identified are reflected in the different criteria used to assign fish to the various consumption categories. For example, in New York, if the sum of the ratios mentioned above exceeded 1.5, then anglers would be advised to eat those fish no more often than one meal a month. New York recommended no consumption when the sum of the ratios exceeded 3.0.

In contrast, Pennsylvania and Ohio simply applied the FDA action/tolerance levels to sport fish. Thus, if the PCB levels were above 2.0 p.p.m., they recom-

Table 13.4 Task Force Survey: similarities among advisory programs of the eight Great Lakes and Province of Canada

---

Interagency coordination procedures exist within each jurisdiction.
All but one jurisdiction in some way apply federal action/tolerance levels as the basis for sportfish
consumption advisory triggers.
All jurisdictions issue special cautions for women and children.
All jurisdictions publicize methods of trimming and cooking fish to reduce exposure to contaminants.
All but one jurisdiction use skin-on fillets as the standard sample preparation method for the primary fish
species analyzed from the Great Lakes.
All but one jurisdiction issue updates on advisories at least annually.

Table 13.5 Unsolved issues among the eight Great Lakes states and the Province of Ontario. Specific criteriafor placement of fish into advisory categories

---

Number of advisory categories and specific consumption advice for each.
How and when risk assessment methodologies should be applied.
How and when to compensate for presence of multiple contaminants.
Best ways to communicate risks and publicize the advisories.
Selection of contaminants to be monitored.
Interlaboratory quality assurance/quality control checks.
Extent to which the finalized EPA guidance manual on 'Assessing Health Risks from Chemically Contaminated Fish and Shellfish' should be followed

---

mended no consumption and if they were less than 2.0, there were no restrictions. As discussed earlier, the Lake Michigan states used the percent of samples exceeding the FDA tolerance level to assign fish to the three categories they selected. Minnesota had yet a different approach which used the limit of detection as the advisory trigger for PCBs and dioxins, i.e., if PCBs or dioxins were detectable, the recommendation would be no more than one meal a month. Minnesota recommended no consumption only for fish containing mercury above 2.81 p.p.m.. Ontario used the Canadian federal action levels that were very comparable with the FDA values. If PCB levels were over 2.0 p.p.m., the fish would be placed in a restrict consumption category. Ontario did not issue any no consumption recommendations based on PCB levels.

This survey showed that, although there were significant differences in protocols, eight of the nine jurisdictions based their advisories to some degree on national action/tolerance levels. This was significant because of criticisms of the use of these levels to promulgate advisories and suggestions that risk-based approaches be considered as an alternative.

13.9   WHY NOT USE FOOD AND DRUG ADMINISTRATION LEVELS?

The Task Force then considered the main arguments against using FDA action or tolerance levels as the bases for sport fish advisories. The first is that fish purchased in the market are assumed to come from a variety of sources; fish consumed by anglers, however, often originate from the same body of water. If that body of water is contaminated, this leads to a continual exposure to contaminants. Secondly, the FDA by law has to consider economic impacts. Therefore, the regulatory levels may not be entirely health-based.

Thirdly, FDA levels are designed to protect the average fish eater nationally and do not take into consideration fish eating habits of local populations that may differ significantly from the national average. For example, fish consumption surveys suggest that the average consumption rate for anglers is about three times higher than the rate for non-anglers.

13.10   DEVELOPMENT OF A NEW PROTOCOL

The Task Force then examined alternatives to the use of FDA action/tolerance levels. One was to base the advisories on the carcinogenic potential of fish consumption as cancer was of great public concern at that time. However, the only quantitative cancer values were those generated by the EPA based on cancer risk extrapolations that contained considerable uncertainties and that the states thought to be overly protective. As a result, the Task Force felt that the cancer extrapolation approach based upon animal data was not a significant improvement compared with the FDA action/tolerance levels approach and that people would be more likely to understand the latter. Thus, the Task Force agreed not to move toward using cancer as the critical health endpoint for advisory development.

Instead, the Task Force decided to develop a 'Health Protection Value (HPV)' for PCBs using an approach similar to the one EPA uses to establish reference doses (RfD) for noncarcinogenic health endpoints. PCBs were selected as the contaminant of concern as PCBs were the contaminants occurring at the highest levels in Great Lakes fish.

To start the HPV development process, the Task Force did an extensive review of the literature and calculated an RfD for PCBs from the data reported in each study using standard risk assessment methodologies, including the application of safety or uncertainty factors. In addition, the strengths of each of these studies were evaluated. However, rather than using a single study as the EPA does in establishing RfDs, the panel examined RfDs generated by a number of organizations as well as itself and decided to use the central tendency of these RfDs. This value, 0.5 μg/kg per day, was the HPV.

To be sure this value was not too permissive, the Task Force calculated a cancer risk estimate corresponding to exposures at the HPV. The result was an excess cancer risk of approximately 1×10^{-4} (1 in 10 000), a value within the range of cancer risks that the EPA considers acceptable (10^{-4}–10^{-7}). In light of this result and the strong support from the regional EPA representatives, the Task Force decided to adopt this HPV, which corresponds to a daily dose of 3.5 μg of PCB per day.

In addition, the Task Force decided that instead of using three consumption categories (unrestricted, restricted, and no consumption), there should be five categories so as to better reflect the continuum of risks from the least contaminated species to the most contaminated species. In this new system, the unrestricted category corresponds to up to 225 meals a year. There are three restricted consumption categories. One is one meal a week, a category that most of the states were already using. The other two are one meal per month and a six meals per year. This last restricted consumption category was included to address risks for seasonal anglers who fish in the Great Lakes only a few times a year. The fifth category is no consumption.

Other assumptions to which the Task Force agreed were that the average fish meal size is about 250 g and the weight of the average adult is 70 kg. It also decided that the literature supported inclusion of a factor of 0.5 to take account of reduction of contaminant levels by trimming and cooking. Based on these assumptions, the panel calculated the PCB concentrations that correspond to each of the advisory categories. Any fish containing PCBs at levels below 0.05 p.p.m. would fall into the unrestricted consumption category. One meal a week corresponded to 0.05–0.2 p.p.m.; one meal a month to 0.2–1.0 p.p.m.; six meals a year to 1.0–2.0 p.p.m.; and above 2.0 p.p.m. to no consumption. It is coincidental that the no consumption level is quite similar to the FDA tolerance level.

Figure 13.5 provides an example of how the proposed Task Force protocol would be implemented for chinook salmon from Lake Michigan based on pooled monitoring data from Michigan, Illinois, and Wisconsin for the years 1988–1991. The plot shows that, with this data set, chinook under 27 inches in length would fall into the one meal a month category and those above 27 inches would be assigned to the six meals a year category. No salmon would be in the no consumption category as the regression line doesn't cross the 2.0 p.p.m. criterion within the species normal size range.

Minnesota liked the Task Force draft approach so well that it adopted it on a trial basis in 1992 and has kept it in place since. To communicate the categories to anglers, Minnesota publishes an advisory booklet that has symbols in each of the size categories corresponding to different levels of contamination; a totally shaded-in fish representing no consumption and a fish without shading representing unlimited consumption (Minnesota Department of Health).

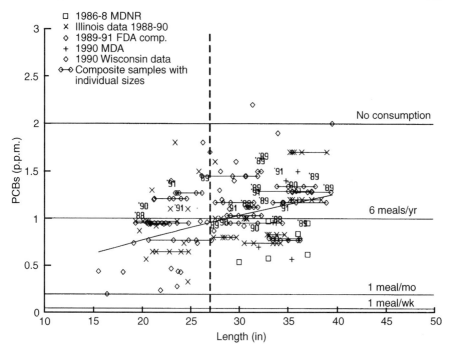

Figure 13.5 Polychlorinated biphenyl (PCB) concentrations versus fish length in chinook salmon from Michigan, Illinois and Wisconsin waters of Lake Michigan, 1988–91, and tentative advisory categories using the Task Force proposed protocol. Fish to the left of the vertical dashed line (less than 27 inches) would be placed in the one meal per month category. Those to the right of the dashed line (27 inches or longer) would be placed in the six meals per year category.

13.11   PEER REVIEW

The results of the Task Force's deliberations were peer reviewed, and the proposed protocol was given to the Great Lakes governors in September 1993 for their consideration. Because this protocol was more restrictive than those in effect in several states, and thus could have significant economic implications, the Council of Great Lakes Governors decided to request an additional peer review by a special science panel.

The panel's report, published in September 1995, concluded that the HPV that the Task Force had selected was appropriate for women of child-bearing age and children, but was probably overly protective for the rest of the population (Fischer *et al.* 1995). It recommended consideration of a separate HPV for the remainder of the population. The panel also recommended that the proposed advisory should more effectively communicate the relative risk of eating the contaminated fish compared with the risks associated with other common foods that people eat.

Another panel recommendation was that the advisory should more strongly emphasize the health benefits of eating fish. The Task Force's proposed common advisory contained a few general statements about health benefits, but the panel felt that these statements did not adequately reflect the extensive studies demonstrating the health benefits of eating fish that are found in the literature.

13.12 CURRENT STATUS

The panel's recommendations were referred back to the Great Lakes Task Force which met in March 1996 and submitted its response to the Governors' Council. This information was transmitted to the governors in each state who, working in conjunction with their health agencies, will make the ultimate decision on the approach to be taken within their jurisdictions.

13.13 CONCLUSIONS

The history of fish consumption advisories in the Great Lakes basin illustrates both the barriers to harmonization and the ways that they can be surmounted, at least to some degree. The barriers include numerous governmental agencies with overlapping responsibilities, as well as political and economic differences among states and provinces and inconsistencies in data collection and interpretation. For a time, these barriers contributed to a situation where advisories in neighboring states for the same fish in the same bodies of water differed significantly.

However, through a stepwise process starting with agreements among the four states bordering Lake Michigan, commonality among advisories issued by all Great Lakes states has greatly increased. In addition to the willingness of these states to pursue harmonization of advisories vigorously, the inclusion of a number of nongovernmental interested parties in the process was instrumental in reducing inconsistencies. While harmonization has increased, inconsistencies have not been eliminated and it appears that political processes in each state will remain a barrier to further movement toward common advisories.

It should also be recognized that the actions at the state level have not been paralleled at the federal level and that there remain differences among federal agencies, namely the EPA and FDA, as to the best approach to take in managing the risks of contaminants in fish sold in interstate commerce. Efforts at the federal level to address these differences will be discussed in Parts II and III of this chapter.

REFERENCES

Fein GG, Jacobson JL, Jacobson SW, Schwartz PM, Dowler JK (1984). Prenatal exposure to polychlorinated biphenyls: Effects on birth size and gestational age. *Journal of Pediatrics* **105**: 315–320.

Fischer LJ, Bolger PM, Carlson GP, Jacobson JL, Knuth BA, Radike MJ, Roberts MA, Thomas PT, Wallace KB, Harrison KG (1995). *Critical Review of a Proposed Uniform Great Lakes Fish Advisory Protocol.* Michigan Environmental Science Board, Lansing, MI.

Hesse JL (1990). Summary and analysis of existing sportfish consumption advisory programs in the Great Lakes basin. A discussion document of the Great Lakes Fish Consumption Advisory Task Force. April 30.

Hovinga ME, Sowers M, Humphrey HEB (1992). Historical changes in serum PCB and DDT levels in an environmentally-exposed cohort. *Archives of Environmental Contamination and Toxicology* **22**: 362–366.

Humphrey HEB (1991). Environmental contaminants and reproductive outcomes. *Health and Environment Digest* **5**(8): 1.

Jacobson SW, Jacobson JL, Schwartz, PM, Fein GG (1983). Intra-uterine exposure of human newborns to PCBs: Measures of exposure. In: PCBs: Human and Environmental Hazards (D'Itri FM and Kamrin MA, eds). Butterworth, Boston, MA. pp. 311–343.

Michigan Department of Public Health (1983). Evaluation of humans exposed to water borne chemicals in the Great Lakes. Final report to the Environmental Protection Agency, Cooperative Agreement CR-807192. Washington, DC.

Minnesota Department of Health (1995). *Minnesota Fish Consumption Advisory.* Division of Environmental Health, St. Paul, Minnesota.

Paneth N (1991). Human reproduction after eating PCB-contaminated fish. Commentary. *Health and Environment Digest* **5**(8): 4.

US Environmental Protection Agency (1995). *National Listing of Fish Consumption Advisories.* EPA 823/C-95/001. Office of Water, Washington, DC.

CASE STUDY COMMENTS: PART I (P. Michael Bolger)

Center for Food Safety and Applied Nutrition, US Food and Drug Administration, Washington, DC, USA

First, it is important to distinguish between an action level and a tolerance. The Food and Drug Administration (FDA) sets action limits but the US Environmental Protection Agency (EPA) Office of Pesticide Programs sets tolerances for pesticides with some input from the FDA. The polychlorinated biphenyl (PCB) level discussed earlier is a tolerance, while the level for mercury is an action level. A tolerance is established based on hazard issues, loss of food considerations, and other factors, e.g., detection limits. It is often stated that action levels are not risk based but this is not the case. They are based on a qualitative consideration of the risk and it is not clear that a quantitative assessment utilizing a reference dose is more appropriate for improving public health. Indeed, there are no data to document many of the action levels so it is difficult to show which approach might be most appropriate.

The situation with respect to the mercury action levels is complex. The FDA established the level at 1.0 p.p.m. rather than 0.5 p.p.m. because the lower value was considered overly conservative and unrealistic. This decision was based in part on new data from mercury-exposed populations in Japan and Sweden suggesting that the dose–response observed there was not applicable to other populations.

Methylmercury is a different issue from PCBs because it is found in almost any fish that is tested. In addition, it appears that most of the methylmercury exposure is from ocean fish so that fish advisories are not likely to reduce risk greatly, particularly from a national perspective. However, it is apparent that much better data need to be collected on methylmercury exposure in anglers before this comparison can be made quantitatively with confidence.

With respect to the PCB-based Health Protective Value (HPV), it is difficult from the Task Force report to determine exactly how it arrived at its final HPV value. Instead of a quantitative rationale being provided, it appears that it resulted from a consensual process. Questions have been raised about such a process because the outcome seems to depend heavily on the composition of the consensus group. This, in turn, can lead to inconsistencies as different expert groups evaluate different contaminants. Thus, it is important that the framework for establishing expert groups and encouraging consensus be explored carefully.

Another issue related to the HPV, or to any safety assessment value, is that it is based on the criterion of zero or negligible risk. This number does not provide any information about the risk that might be incurred if the value is exceeded. In addition, the HPV approach does not provide the consumer with useful quantitative information about nutritional issues or competing dietary risks. Thus, anglers do not have all the information they need to make the best decisions about fish consumption. Consideration needs to be given to the best ways of quantitatively incorporating all this information into fish advisories or to other ways of educating the public about contaminants in fish.

CASE STUDY COMMENTS: PART 2 (William H. Farland)

National Center for Environmental Assessment, US Environmental Protection Agency, Washington, DC, USA

First, it is important to recognize that fish advisories are a national issue. Currently, 46 states have fish advisories, although seven are single statewide advisories. These advisories cover about 14% of the total lake acreage and about 4% of the total rivers in the United States and include fish from the oceans as well as the inland waters. Thus, advisories pose a very significant harmonization issue.

In 1990, the US Environmental Protection Agency (EPA) decided to develop a federal action plan for fish advisory activities in response to state requests for a number of federal actions. The first recommendation was to develop consistent EPA and Food and Drug Administration (FDA) risk assessment approaches, including a decision as to whether the action level or some risk-based number is the most appropriate criterion for establishing advisories. In addition, states wanted guidance in developing consistent risk-based methods for use in fish advisories as well as federal support for fish consumption rate surveys that could supply information needed in promulgating advisories. Another issue of concern to the states was federal government funding for monitoring and sampling programs to determine temporal changes in contaminant levels in fish. In addition, states saw a need for training workshops and a clearinghouse for fish advisory related information.

The action plan developed in response to state concerns includes programs to improve the information exchange between the state and federal programs, and to improve national databases to support fish advisory development. As part of the plan, the EPA has undertaken some fairly aggressive activities, including the issuance of guidance volumes for fish sampling and analysis and for risk assessment related to fish consumption. A risk management guidance document was issued recently, and a risk communication document is in development. In addition, the EPA has sponsored national technical conferences dealing with fish advisory issues and has developed and distributed national fish contamination and fish advisory databases. The EPA also organized a technical meeting on advisories, in June 1996, that included representatives from 47 or 48 states.

Unfortunately, there has been little progress on the first recommendation with regard to harmonizing EPA and FDA approaches and thus, little guidance is available to the states on this issue. Also, there has been limited progress with regard to fish consumption rate surveys, with efforts focused on some American Indian tribes and on providing some funding for technical assistance and monitoring to the states. Another of the recommendations, to address comparative risk, has not been dealt with, so there is quite a bit left to do on these issues.

Some progress is being made on the question of whether or not to change the way that the science is used in fish advisories. For example, there is an attempt to eliminate the dichotomy between cancer and noncancer endpoints, and to deal with the issue of high versus low levels of exposure in these fish advisories. The blood data that Mr. Hesse (see p. 154–6) provided suggested that with certain levels of consumption an individual might have eight or more times the average blood levels for a particular contaminant. This suggests that some action is needed to prevent such levels from occurring. However, before deciding on the most appropriate actions, it is important to determine whether or not fish consumption is the major source for those contaminants in humans and to identify and quantify other possible sources of human exposure to these contaminants in the environment.

The scientific issues of better understanding of background and cumulative exposures, dose–response for both carcinogens and noncarcinogens, and whether or not some cancer endpoints have a threshold for response are issues that need to be grappled with. Finally, in terms of risk characterization, the question of how exposures from sport-caught fish relate to action levels for contaminants in commercial fish is something that needs to be addressed.

14

Workshop Summaries

MICHAEL A. KAMRIN
Institute for Environmental Toxicology, Michigan State University,
East Lansing, MI, USA

14.1 EXPOSURE ASSESSMENT WORKSHOP

Michael A. Kamrin, Facilitator

14.1.1 INTRODUCTION

Exposure assessment is the part of risk assessment that involves the determination of the amount, frequency, duration, and route of exposures of populations to environmental chemicals. This determination requires information about the concentrations of the chemical in various media and the behaviors of potentially exposed populations. Environmental concentration data are usually gathered by sampling and analysis performed at or near the site of interest, while behavioral parameters are generally based on generic studies of populations in a variety of locations.

There are a number of areas where inconsistencies in exposure assessment may arise. One is in the determination of environmental concentrations. Different laboratories may use different analytical techniques resulting in different analytical results. Sampling schemes may also have a great influence on both the identities and concentrations of contaminants detected. Further, questions of how to account for samples that are below the limit of detection can lead to disparities in exposure assessments.

When exposure concentrations are estimated using transport and fate models, the situation is more complex, thus increasing the possibility of inconsistencies. Nonetheless, such models are often needed to estimate the time course of environmental concentrations at some distance from a source. For example, hydrogeological models are used to calculate concentrations in drinking water downgradient from a contaminated site.

The selection of behavioral parameters can also lead to inconsistencies from

Environmental Risk Harmonization: Federal and State Approaches to Environmental Hazards in the USA.
Edited by M. A. Kamrin. Published in 1997 by John Wiley & Sons Ltd, Chichester. ISBN 0 471 97265 7

one assessment to another. While the US Environmental Protection Agency (EPA) provides extensive guidance as to the most appropriate assumptions for a variety of behavioral parameters, these may not always be applicable to the situation of concern (US EPA 1989). For example, there may be particular subpopulations at risk whose behaviors are significantly different from those upon which the default values are based. Often, valid data about these subpopulations are limited and different risk assessors may legitimately make different assumptions about their behaviors.

Inconsistencies may also arise because of the choice of exposure routes included in the analysis. For example, the current drinking water standards are based only on ingestion while it is clear that the presence of volatile chemicals in drinking water may also lead to dermal and inhalation exposures through washing, showering, and perhaps other activities. In addition, surveys indicate that there is significant variation in the routes of exposure to soil contaminants that are considered in risk assessments performed by individual states (National Governors' Association 1994).

14.1.2 WORKSHOP DISCUSSIONS

The members of the workshop group discussed a number of these disparities and focused on several issues that are significant to harmonization from the state perspective. One is the availability of data and analyses that could assist in developing parameters for exposure routes that are not now addressed and in improving parameters for exposure routes that are now taken into account. For example, information about dermal absorption and bioavailability of chemicals in soil is very limited and this makes it difficult to incorporate this route in exposure assessments of soil contaminants.

Another type of information that is needed is how best to assess the distribution and fate of chemicals in the environment. This kind of information is necessary for estimating exposures from distant sources or for estimating the time course of exposure. A number of models are currently in use but none have been validated well enough to be considered the model of choice. As a result different states have adopted different models, another factor contributing to disharmonies in exposure assessment.

Another issue is the impact of local and regional differences on exposure parameters. While some exposure parameters are likely to be site independent, e.g., bioavailability, others such as water consumption may vary regionally and not always correspond to national default values. As a result, the use of such defaults may lead to exposure assessments that overestimate exposure in some parts of the country and underestimate it in others.

The participants also suggested that decisions as to exactly what segment of the population to protect and how to report exposure assessments provide opportunities for inconsistencies. In particular, the reporting of a single exposure value protective of an undefined segment of the population is being replaced in some places by the reporting of a distribution of population exposures for the whole population. Utilizing the latter to decide the most appropriate exposure values to use in risk assessments can lead to much different exposure numbers than those generated from point estimates for a hypothetical population group.

In addition to these technical issues, the amount of commonality that can be achieved among states and between states and the federal government is limited by the increasingly prescriptive legislation being enacted at the federal and state levels. In Michigan, for example, the requirement that risks at hazardous waste sites be assessed on the basis of future land use has led to the development of corresponding exposure assessment approaches (see Chapter 10). The land-use approach is in distinction to previous Michigan legislation, current legislation in most other states, and current EPA policy that requires utilizing the assumption that all sites will eventually be used residentially.

A related issue is intrastate inconsistencies that may arise when legislation governing different media prescribe different exposure assessment approaches in different agencies or offices within agencies in the same state. Again using Michigan as an example, the corresponding exposure assumptions used by units that address hazardous wastes may not be the same as those applied by agencies or divisions dealing with risks from surface water or ambient air.

14.1.3 WORKSHOP RECOMMENDATIONS

The workshop participants felt that a number of steps might profitably be taken to reduce inconsistencies in the assessment of exposures to environmental contaminants. As indicated in the discussion, these steps are limited by the legislative mandates and administrative procedures that are currently in place in the various states and at the federal level. However, even within these limitations significant actions can be taken.

Some of these should be the responsibility of the federal government; either to perform or to support financially. One is to provide additional data and analyses needed to better estimate generic exposure parameters, such as soil absorption and bioavailability. A second is the development and validation of transport and fate models that can provide the most accurate estimates of environmental concentrations remote from sources both spatially and temporally.

Others are more appropriately the responsibility of the states, acting separately or together. These involve careful examination of local and regional conditions that may require modifications of the EPA default values, or soil guidelines. These factors may include meteorological conditions, variations in population composition and unique behaviors of populations or subpopulations. An example of a meteorological factor is snow cover limiting soil contact; an example of variations in population composition is age distribution; and an example of unique behaviors is a subpopulation subsisting on a specific food source.

In some cases, the steps taken by a state or region can serve as a model for a national approach. The case study of the Great Lakes fish advisory is a good example of this (see Chapter 13). The states in the region came together to develop a common approach to exposure and risk and this has served as the impetus for a federal harmonization effort including all of the states. It is important to note that this and other examples discussed previously make a good case for greater interstate cooperation in risk assessment and risk management, including exposure assessment.

In general, the workshop participants concluded that harmony in exposure assessment implies that everyone will work together within a common framework based on commonly accepted scientific data and models. States may come to different absolute exposure values because of differences in legislative mandates or differences in exposure conditions. However, changing the legislative framework was considered outside of the purview of risk assessors and differences in exposure conditions were considered justifiable grounds for disparities from state to state.

14.2   TOXICITY ASSESSMENT WORKSHOP

Michael L. Dourson, Facilitator

14.2.1   INTRODUCTION

As used in human health risk assessment, toxicity assessment is a complex activity with the goal of predicting the doses at which specific chemicals will cause particular adverse effects. The results of toxicity assessments are generally combined with exposure assessments in a process called risk characterization, which is the end product of the risk assessment process. The characterization provides a description of the risk from a particular chemical or group of chemicals in a particular exposure situation, e.g., gases emitted from a municipal waste incinerator.

For historical reasons, toxicity assessments are performed differently to assess carcinogenic as compared with noncarcinogenic effects. This distinction is largely based on the assumption that carcinogenic effects do not have a threshold so that the assessment cannot estimate a safe level; instead it is an estimate of the cancer incidence that will result from daily lifetime exposures to a particular dose of a specific chemical. Toxicity assessments for carcinogens are generally performed, at least in the United States, by extrapolating high-dose animal data to much lower environmentally relevant doses using mathematical models. Interspecies conversion factors are then used to develop toxicity values applicable to humans.

On the other hand, as noncarcinogenic effects are considered to have a threshold, toxicity assessments for noncarcinogens are performed with the goal of identifying a level below which adverse effects are not expected to occur. These assessments are generally based on studies of experimental animals and utilize either the highest dose at which no adverse effect is observed (NOAEL) or the lowest dose at which an adverse effect is observed (LOAEL) as the basis for calculating safe values. The NOAEL or LOAEL is divided by a series of uncertainty or safety factors to account for inter- and intraspecies variability, data quality, and other factors to arrive at an acceptable daily exposure value. Some examples of such values are the ADI (acceptable daily intake), RfD (reference dose), TDI (tolerable daily intake), and the MRL (maximum risk level).

The above description does not include mention of differences in approaches to assessing carcinogenic and noncarcinogenic effects that are either in use or being considered in the United States and abroad. These include the use of safety factors for certain types of carcinogens and the use of a benchmark dose (representing a particular percent incidence for an adverse effect) for both carcinogens and noncarcinogens. In addition, there is currently much discussion about incorporation of additional information, e.g., mechanism of action or toxicokinetics, in extrapolating data from one dose to another and from animals to humans.

14.2.2   WORKSHOP DISCUSSIONS

Considering the amount of variation in practice and under discussion, it is not surprising that the workshop participants had disagreements about some fundamental issues in toxicity assessment, although they did agree that harmonization of toxicity values was an appropriate overarching goal. For example, some thought that the NOAEL or LOAEL represented toxicity values but that the ADI did not. Others disagreed and felt that the latter were also toxicity values. This pointed up the importance of clear definitions to the harmonization process.

Further, the importance of scientific judgement in toxicity assessments, whatever the approach, suggested that standardization of toxicity values would be difficult. The workshop participants felt that having standardization as a goal was not appropriate; rather consistency in the risk assessment process would be a better aim. In addition, they suggested that improved communication among assessors and better incorporation of scientific knowledge by assessment groups, e.g., agencies, should be considered measures of success in harmonization. There was also a consensus that harmonization should be occurring among and between parties at all levels—local, national, and international—at the same time. Groups at each level should not wait for the others to act first.

The discussants also considered current barriers to harmonization in risk assessment and were able to identify several. One is liability concerns that has led some industry scientists to be very wary of documenting and making available risk assessments they have performed. A second is political where assessments may be strongly influenced by policy considerations. Related to this is the impact of economic considerations, e.g., where chemicals of great economic importance may be assessed differently from those of lesser import. Another barrier, also mentioned in the presentations, is prescriptive legislative mandates that delimit the approaches that can be adopted. Other barriers discussed included the related problems of assessment information dissemination and lack of understanding of the assessment process. It was noted that there are few, if any, effective mechanisms for exchanges of toxicity assessment information among assessors and between assessors and stakeholder groups.

14.2.3 WORKSHOP RECOMMENDATIONS

In light of the above discussion, the workshop participants made a number of recommendations. One is to clarify the process of toxicity assessment. This can be done by first identifying current accepted practice(s) and carefully examining the elements of this practice. The next step is to develop a strategy that provides the greatest potential for harmonization. The attendees recommended that discussants start from less controversial principles, such as the use of peer review, and move on to more contentious issues. During this process, opportunities for improving current practice should be identified and encouraged. This recommendation is made with the understanding that toxicity assessment should be thought of as a dynamic process that needs to be under constant review rather than as an unchanging set of prescriptions.

Another recommendation was to increase the uniformity and understandability of the reporting of toxicity assessment values. In this way, even if different values are considered best by different assessors, those who might use these values will understand the bases for the differences and thus which values might be most appropriate for their applications. To enhance communication, it was proposed that toxicity assessment information be collected and organized at one site, perhaps an Internet web site, where it can be readily accessed and where information can be changed readily to reflect new inputs.

14.3 RISK MANAGEMENT WORKSHOP

William E. Cooper, Facilitator

14.3.1 INTRODUCTION

Risk management is generally defined as the actions taken to minimize or eliminate risks to human health and the environment. Management strategies are based on risk assessment results combined with social, economic, political, and technical factors. In the ideal situation, risk assessments are performed independently of these latter factors. However, in practice, it is clear that risk management considerations are integral to the risk assessment process. For example, the use of the most sensitive species or the application of certain safety factors in human health risk assessment represents a management decision and not a scientific one. From a scientific standpoint, the species most applicable to humans is the best choice and any safety factors should be based on knowledge of the biochemistry and physiologies of the test animal as compared with humans.

As a result of historical factors, risk management of chemicals in different environmental media is carried out under separate pieces of legislation that may mandate different risk management approaches. For example, the regulation of a contaminant in fish may explicitly include economic considerations, while the regulation of the same contaminant in drinking water may not. Such differences are common in legislation at the state as well as the federal level.

The media specificity of environmental statutes and rules as well as political, economic, and social factors have led to significant inconsistencies in the risk management of the same chemicals in different media and even the same chemicals in the same media in different places. For example, cleanup requirements for the same chemical in soil at hazardous waste sites in two different states may be quite different reflecting different priorities and values in these two locations.

14.3.2   WORKSHOP DISCUSSIONS

The workshop participants discussed the amount of harmonization in risk management that was desirable and also practically achievable. They did this in the context of federal and state approaches that are currently in use and also in light of changes that are occurring or proposed. In addition, they examined steps that might profitably be taken to increase consistency among and between federal and state agencies, when desirable.

With respect to the fundamental question of whether or not a common, unified federal statute across all media was desirable or not, the group concluded that it was not. However, the group felt that the goal of such a unified statute, i.e., multimedia risk management, was very desirable. The group opinion was that risk management was still evolving and that there was not enough consensus to serve as the basis for a common statute. In the current situation, flexibility is important so that experimentation with different approaches can occur. In addition, there was a concern that opening up all of the current statutes for discussion could lead to a lessening of overall environmental protection.

The workshop participants also concluded that a unified state statute was not yet appropriate for the same reasons as cited with respect to a federal statute.

However, it might be appropriate to combine parts of current legislation where it is clear that such combinations would lead to more effective risk management. The example of an agency staff member trying to address all of the different media impacts of a hazardous waste site illustrates a situation where it would be advantageous to be able to address several media within one administrative framework.

The Michigan attempt to write a unified environmental code illustrated well what might be accomplished in revising environmental regulations. The commission that performed this work used a functional taxonomy to develop a totally different framework from the current media-specific one. This framework separated natural resource concerns from the environmental ones and, within the environmental area, point sources from nonpoint sources. The strategy for all types of point sources in all media might be similar, e.g., command and control, while management of all types of nonpoint sources in all media might also be similar but different from that for point sources, e.g., market incentives. This framework was accepted by the Michigan legislature, although some specific issues remain to be resolved.

The interstate harmonization of risk management was considered infeasible in light of the social, economic, and political differences among states. However, as suggested by actions in the Great Lakes region with respect to health surveys and fish consumption advisories, greater coordination in risk management among states is desirable and achievable. This is especially important when there is a shared resource such as a watershed, river, lake, or shoreline. The participants also felt that the federal government, especially the EPA, should take a stronger role in facilitating coordination among the states.

The workgroup members also suggested that federal–state coordination was both desirable and feasible. The best example is the common framework developed through the American Society for Testing and Materials (ASTM) for making decisions at the state and federal level about remediation of leaking underground storage tanks (ASTM 1995). The Risk-Based Corrective Action (RBCA) process is not proscriptive and allows states to incorporate their own values in applying this framework. For example, different states may set their own acceptable risk levels for carcinogens or decide at what distance from a site compliance with groundwater standards should be assessed.

14.3.3 WORKSHOP RECOMMENDATIONS

The participants recommended that the goal of risk harmonization across and among federal and state programs should be based on a set of common processes and/or goals and not a set of prescriptions of exactly how to perform risk management. As part of this, risk managers should be asked to consider a number of common questions in arriving at their decisions. One is what population groups and/or subgroups the risk management strategy is designed to protect. Protecting 100% of the population is often impractical, and even when it is possible it may lead to the loss of a valuable resource. For example, banning fish consumption because of contaminants could protect 100% of the population but would also mean that the benefits of fish consumption would be lost.

Another question that should always be considered by risk managers is the basis for their decision. Although some regulations are based entirely on risk, almost all real world situations require consideration of factors other than risk. The above cited fish consumption situation illustrates the use of benefit as well as risk in decision making. In other cases, both costs and benefits are taken into account. In any event, this is an issue that the risk manager needs to address explicitly in the decision-making process.

Because of the difficulty of one person being able to appreciate all of the factors that may impact upon a risk situation, risk managers should include a representative cross-section of stakeholders in the decision process. Not only can they provide valuable information about relevant benefits and risks, they may be able to suggest risk management strategies that would be most cost-effective for the groups that they represent.

The group also discussed the impact of the recent federal and state risk comparison efforts (US EPA 1987; Michigan Department of Natural Resources 1992; California EPA 1994) on risk management and recommended that care be taken that this type of approach doesn't lead to the establishment and implementation of new priorities at the expense of the old. For example, as indoor air becomes a higher priority and receives increased funding, this should not be accompanied by a decrease in the rigor of drinking water or hazardous waste site risk management actions.

One other issue was the impact of industry actions in the area of risk management, e.g., ISO 14000 standards. These standards, which are under development by the International Standards Organization, provide a common environmental management system for industries in different countries. The goal is to provide a common mechanism for satisfying the environmental standards of nations around the globe. While this may have some impact on risk management decisions in the coming decades, the group suggested that these steps must be taken with due consideration to the broader risk management context.

14.4   RISK COMMUNICATION WORKSHOP

Carol Y. Swinehart, Facilitator

14.4.1   INTRODUCTION

Risk communication is the process by which the results of risk assessments and risk management decisions are communicated to a variety of audiences. These include the public, industry, environmental and public interest groups, government agencies, and scientists. While this has often been thought of as a one-way process where information flows from the assessor and manager to the audience, the more desirable approach, as articulated in the National Academy of Sciences report entitled *Improving Risk Communication* (National Research Council 1989), is that this become a two-way process with the audience participating in the message formulation as well.

The question of harmonization in risk communication can be thought of in two different ways. One is how well the process of risk assessment and risk management is being communicated. This has a significant impact on the public's perceptions of the existence of and justification for disparities in the assessment and management of environmental risks. The other question is whether risk is being communicated in a harmonious way by the various actors involved in risk assessment and risk communication. A lack of coherence can have a strong negative impact on credibility and thus the ability of agencies to move from current approaches to new ones that may be more consistent with those in other units. The problems implicit in these questions are probably exacerbated by the significant scientific and methodological uncertainties that are inherent in risk analysis.

14.4.2 WORKSHOP DISCUSSIONS

The group first addressed a number of problems that tend to reduce harmony in risk messages. The first is that consistent messages are not always provided, even by representatives of the same level of government. For example, environmental and public health agencies may release conflicting information about the severity of particular environmental risks. The second is that there may be confusion about the purpose of the communication; is it educational or designed to bring about attitudinal or behavioral change? Audiences may be skeptical of messages that are described as educational but clearly are persuasional, e.g., that exaggerate the risk. Another problem is presenting the risk in an appropriate context. Comparisons between voluntary and involuntary risks may not be accepted if perceived by audiences as a way of minimizing the importance of the involuntary risks.

Another barrier to harmonization may occur when the communication is one-way and does not include feedback at appropriate stages in the process. If the key stakeholders are not involved at the beginning of the process their input may be limited and the chances for harmony greatly reduced. In addition, if the communication is not a continuing process, but limited to a few events, there is less chance for meaningful risk communication.

Another problem that often arises is that the message is not tailored to the audience. As a result audiences may be presented with material at levels of complexity that are inappropriate or in contexts to which they cannot relate. The problem may also become more difficult when there are multiple audience segments and therefore different strategies that are appropriate for each. Intermediaries, e.g., journalists, represent an audience of particular importance. It may be difficult to communicate effectively with reporters because of their varied backgrounds and because of the necessity for rapid and accurate response.

Another critical issue is how to communicate uncertainty, which is such an important element of risk assessment and risk management. If the uncertainty is not explained adequately, the audience does not have enough information to make an informed contribution to risk resolution. In addition, audience participation may be limited if choices for dealing with this uncertainty are not presented as part of the risk communication process. As a result, risk management outcomes may be ineffective or less than optimal.

14.4.3 WORKSHOP RECOMMENDATIONS

The first recommendation is that risk assessors and risk managers should work together to provide consistent messages to the extent possible. The meetings of this and the other workshop groups represent one example of this type of effort. The second recommendation is that the goal of the communication be clear; educational, or persuasional. The aim may be different depending on the risk management option that is chosen. Third, risk comparisons should be used to communicate risk but care must be taken that they are appropriate to the situation and the audience.

When dealing with uncertainty in risk assessment and management, it is important that both the nature and the sources of uncertainty be explained as well as the implications of this uncertainty for risk decisions. It is also helpful for the communicator to give the audience his or her own personal decision and the reasons for it and to leave the audience members to make their own decisions. To accomplish this last goal, it is important to provide simple advice that allows for meaningful choice, e.g., eat smaller rather than larger fish.

When communicating with journalists, they should be thought of as concerned citizens who need assistance in understanding the context of the situation as well as the current events. They should be responded to even on short deadlines, but should also be encouraged to do follow-up stories in more depth. It is a good idea to interact with reporters outside of crisis situations, e.g., in workshops, to establish relationships with these individuals.

Risk communicators should also take advantage of other intermediaries that are trusted by the public, such as physicians. Citizens should be encouraged to discuss their concerns about environmental risks with physicians or other health professionals. They can do this through direct contact with local health providers or by accessing hospital health education programs. Short publications aimed at lay audiences but made available to physicians, e.g., tearout sheets in medical journals, can also be effective communication tools.

School children can also serve as conduits for risk information. Developing and distributing curriculum materials on risk issues and presentations in classes or to teachers are ways that risk communicators can indirectly reach a broad audience of adults. Children can also be reached through the mass media as described previously. A more recent and potentially powerful way to reach youngsters is through the Internet.

Risk communication that promotes harmony requires an appropriate forum and the participation of multiple stakeholders. For example, at both the state and federal level, a coordinating group of agency representatives should provide a forum to interpret risk assessment results and share communication strategies. Public involvement should be a critical element of this process. It is also important to have clear leadership of the forums. One agency can be agreed upon to be the leader or the leadership role can be rotated among the agencies that are involved in the risk assessment and risk management activities.

To accomplish the above goals, it is critical that significant resources be devoted to risk communication. At present, this is not the case and, as a result, a variety of disharmonies have occurred and, perhaps more importantly, the credibility of government agencies has suffered. It is the strong recommendation of the workshop participants that there should be a large increase in the resources, particularly dollars, that are devoted to improving risk communication at all levels of government.

REFERENCES

American Society for Testing and Materials (1995). *Standard Guide for Risk-based Corrective Action Applied at Petroleum-release Sites.* E1739-95. ASTM, West Conshohocken, PA.

California Environmental Protection Agency (1994). *Toward the 21st Century: Planning for the Protection of California's Environment.* California Comparative Risk Project Final Report. Sacramento, CA.

Michigan Department of Natural Resources (1992). *Michigan's Environment and Relative Risk.* Lansing, MI.

National Governors' Association (1994). Risk in Environmental Decisionmaking: A State Perspective. (Working Paper) Denver, CO.

National Research Council (1989). *Improving Risk Communication.* Committee on Risk Perception and Communication, National Academy Press, Washington, DC.

US Environmental Protection Agency (1987). *Unfinished Business: A Comparative Assessment of Environmental Problems.* PB88-127048. Office of Policy, Planning and Evaluation, Washington, DC.

US Environmental Protection Agency (1989). *Exposure Factors Handbook.* EPA/600/8-89/043. Office of Health and Environmental Assessment, Washington, DC.

15

Summary and Conclusions

MICHAEL A. KAMRIN

*Institute for Environmental Toxicology, Michigan State University,
East Lansing, MI, USA*

15.1 INTRODUCTION

This is the first book that explicitly addresses the harmonization of risk assessment and risk management among states in the United States and between states and the federal government, an issue of significant and increasing concern. The importance of this issue is reflected in the contributions of each of the authors and the participants in the workshop discussions. Although the focus is primarily on the United States, the issue is also of international concern as evidenced by the discussions underway among the nations in the European Community and among the current and potential signatories of the North American Free Trade Agreement.

These discussions are occurring at a time when traditional risk assessment and risk management practices are being questioned on scientific as well as other grounds and a number of groups have issued reports on how best to change these practices. This re-examination of currently accepted practices provides an opportunity for incorporating revisions that are responsive to harmonization concerns and this is reflected in many of the suggestions and recommendations made by the contributors as well as by the discussions of harmonization issues in the aforementioned reports.

While a casual reading of each chapter may lead the reader to focus on the many inconsistencies in specific risk assessment and management decisions made by agencies at both the federal and state levels, a more careful examination reveals that there is much common ground upon which to build more harmonized approaches. Building from this common ground starts from a recognition of the factors that contribute to current inconsistencies and examination of approaches that may mitigate these factors.

Environmental Risk Harmonization: Federal and State Approaches to Environmental Hazards in the USA.
Edited by M. A. Kamrin. Published in 1997 by John Wiley & Sons Ltd, Chichester. ISBN 0 471 97265 7

In addition, a number of contributors call attention to the importance of incorporating diverse stakeholders into the risk assessment/risk management process to help in achieving consensus. While it might appear that the addition of a variety of stakeholder viewpoints would increase the potential for inconsistency, the opposite appears to be the outcome. With so many competing viewpoints, it becomes necessary for discussants to focus on a limited number of areas of agreement for progress to be made and for consensus documents to be produced.

15.2 BARRIERS TO HARMONIZATION

The chapters in this book provide testimony to the number and range of inconsistencies that have arisen with respect to the assessment and management of environmental risks to human health. In many cases, these are a direct result of the provisions of the legislation that established and governs the activities of a particular government agency that performs risk assessment or risk management. In such situations, the legislation is a significant impediment to totally common approaches and cannot be completely overcome without new or amended legislation.

In other cases, the regulations that have been developed to carry out the legislative intent reflect the various philosophies of different government agencies with different administrative histories. For example, the Food and Drug Administration and the Environmental Protection Agency have responsibility for assessing and managing risks from chemicals in different media, e.g., food and water. Their divergent approaches have led them to different assessment values in some cases. For example, they both use exactly the same experimental data but often derive different cancer potency estimates. Recent meetings between representatives of these two agencies have led to an agreement on how to resolve this difference, although it is not yet completely implemented. This suggests that such regulatory inconsistencies at the federal level are amenable to resolution, although not without considerable effort in some cases.

The situation at the state level is much more complex because there are 50 states and thus, in aggregate, a correspondingly larger number of government units to bring into agreement. As was clear from the presentations, there are often inconsistencies in regulations among agencies within states as well as inconsistencies among states (see Appendices E, F, and G). It is not clear if harmonization is best achieved through agreement among similar agencies in different states or by consensus among different agencies in the same state. Both strategies are currently being applied; an example of the first is the Federal–State Toxicology and Risk Analysis Committee process (see Chapter 12) and an example of the second is the outside assessment of the California Environmental Protection Agency (see Chapter 11).

Another barrier to harmonization may arise from the lack of a common understanding of the effects being assessed and managed. For example, there is currently a strong push to include endocrine disruption as an effect of concern despite the absence of any validated test for quantitatively assessing this effect. It may well be that there are legitimate scientific differences about the best way to assess this endpoint and these may be reflected in disharmonies in risk assessment and management in different agencies. In this situation, consistency may not be possible or appropriate until the scientific issues are better resolved.

A related problem is that science often moves more rapidly than the legislative process. Thus, one set of regulations may have been promulgated several years in the past and another more recently. Both may reflect science that was current at the time they were issued but, if the science has changed, they may result in different assessments and management decisions. These issues are not easy to resolve but this is an area where further study and thought may result in a regulatory approach that provides enough flexibility to incorporate new science without also requiring lengthy reviews and rulemaking processes.

Another potential barrier to harmonization is the incorporation of different stakeholders in the processes that address different risk issues. The inclusion of such stakeholders is an important part of the recommendations of a number of reports that have been issued in the past few years (see Appendices B, D, and E). More thought probably needs to be given to harmonization problems that may arise from the selection of different stakeholders even for situations which may involve overlapping risk assessment and risk management issues. For example, if one set of stakeholders addresses risks from a hazardous waste site in a community and another addresses risks from an incinerator or industrial facility, the result may be to make decisions that do not best utilize resources to protect human health. This can happen if one group process leads to very stringent controls on a chemical from a minor source while another group interaction leads to a decision allowing much larger exposures to the same chemicals from a major source.

Internationally, trade issues are currently a source of disharmony, although they ultimately may lead to greater consistency among nations in risk assessment and management (see Appendix A). Because of their differences in economic, political, and social structures, different nations approach environmental risks differently. For example, risks from environmental chemicals are likely to be of lower priority in a country that places greater importance on risks that have a greater impact on health outcomes, e.g., nutritional deficiencies. Similarly, different nations may have different qualitative priorities, e.g., chemicals that cause birth defects should be managed more stringently than those that cause cancer. Deep cultural differences are not likely to be overcome easily but, again, imaginative solutions that incorporate local concerns may provide a way to increase consistency among nations.

15.3 THE FUTURE OF HARMONIZATION

The contributors to this book, while recognizing the many barriers to harmonization mentioned above, have pointed out a number of areas where significant progress has been made and where such efforts are continuing and, in some cases, expanding. In addition, and perhaps equally important, they have provided specific illustrations of approaches that seem to be successful and others that have not been as successful.

To start with, all parties recognized that to move ahead it is critical to define what is meant by harmonization. The consensus was that harmonization could not be equated with standardization, although it might involve standardization of some facets of risk assessment or management, e.g., how to determine a reference dose. Instead, the participants generally agreed that consistency was a better synonym for harmonization. Consistency can be thought of as agreement on the general principles of risk assessment and risk management without necessarily agreeing on all of the details.

For example, it might be possible to agree that assessment of exposure to a contaminant from soil should involve consideration of ingestion, inhalation, and dermal pathways and even that particular algorithms are the most appropriate for quantitatively describing each pathway. However, this does not necessarily mean that everyone has to agree on only one default value for each parameter in these algorithms. Instead, these values may reflect both the science and the social judgements of those involved in using this assessment to address a particular environmental risk.

Moving beyond the definition, the message that seemed most clear from discussion of general principles and examination of specific examples is that increased consistency can be achieved only if common frameworks for risk assessment and risk management can be agreed upon and if these frameworks are not overly proscriptive. Frameworks that promote harmonization will have a number of important characteristics, the most critical of which is that they provide for flexibility to suit the circumstances in each jurisdiction, e.g., state and/or nation.

While it is possible to imagine harmonization as agreement on particular risk assessment and/or management methodologies, such consensus is likely to mask existing differences in approach. This, in turn, will often lead to disharmonies in assessment and management of chemicals for which agreement has not been reached, including compounds that may be of particular concern to only some governments. However, if common approaches, rather than common limits, are agreed upon then consistency in risk assessment values and regulatory limits for all chemicals in all media is much more likely.

Agreement was also reached on other desirable characteristics of risk assessment and risk management frameworks. For example, a consensus position was that, considering the limits on available resources, environmental risks should be approached in a tiered, or iterative, manner to address the most significant risks first. If a government unit is faced with a large number of contaminated sites the first step should be to determine which site is clearly of little concern and which is of obvious great concern. Such a screening approach could be done without applying significant risk resources by, for example, using very conservative default values for performing a crude assessment.

Sites that were calculated to be of very low risk even under these very conservative assumptions could then be placed low on the priority list and available resources devoted to refined risk assessments of other sites of greater concern. These more refined, but not complete risk assessments, could take into account more of the characteristics of the specific site and so provide more realistic risk estimates. Based on the priorities established through these assessments, full risk assessments would be performed on the remaining sites. This would provide the best information about the risks associated with these sites, information that could be used not only to compare risks among sites but also between waste sites and other sources of these same contaminants, e.g., indoor air. In addition, the full risk assessment would identify those exposure pathways of most concern and so assist in the selection of the most appropriate remedial options.

Related to this was the suggestion of one contributor to use sensitivity analysis to assess the relative contributions to the risk of the various exposure pathways (see Chapter 8). In the example that was discussed, one that involved assessment and management of a biological risk, it was clear that one of the pathways contributed almost all of the risk and that a risk management approach that minimized this pathway would not only be effective but would also require limited resources. In this way, resources would be freed to deal with other, more significant risks.

Another area of agreement was that the current approach of independently assessing the risk of each contaminant in each environmental medium was no longer the most appropriate. Instead, it was agreed that the risk assessment and risk management framework should incorporate a multimedia approach since this provided the opportunity to: (1) compare risks from the same chemical in various environmental compartments, and (2) combine these, where appropriate, to provide a better assessment of overall risk from that contaminant. Indeed, it was suggested that this approach could be expanded to address groups of chemicals that act by the same mechanism of action. This strategy provides more information than is generally available when the media-specific approach is applied as well as providing the information in a broader context. This additional information and context can only improve both risk communication and risk management decisions.

Some agreement was also reached on the most appropriate relationship between the state and federal governments in assessing and managing environmental risks within this framework. Considering the greater resources of the federal government, it was considered the best source of fundamental scientific data upon which assessments can be based. This includes toxicological information on the levels at which effects in humans might be expected to occur; e.g., the reference dose. It also includes data on human absorption of chemicals through particular routes of exposure, e.g., from skin contact with soil. In the future, it might include information about distributions of behavioral or toxicological variables that could be used in Monte Carlo simulations.

The states' role would then be to utilize this information in combination with knowledge of local environmental conditions and local social values in arriving at the most appropriate risk assessments and risk management decisions. For example, based on meteorological conditions a state might utilize a soil exposure value that was different from a general population default. In another case, the state might decide to address risks that might be quantitatively smaller but considered of greater potential severity by the populace.

15.4 MECHANISMS FOR FUTURE HARMONIZATION EFFORTS

While it is possible to reach agreement on harmonization goals, it is often difficult to identify the best strategies to implement this consensus. As was pointed out in the preceding chapters, there have been a number of harmonization efforts in the United States at both the state and federal levels. Unfortunately, most of these have been relatively short-lived and have not survived changes in political philosophies that resulted from new leaders being elected.

Perhaps, the factors that have brought harmonization issues to the forefront will also result in increased commitments to promoting consistency in risk assessment and risk management at all levels of government, both in the United States and elsewhere. The contributors to this book have provided several examples of approaches that have worked and suggestions of mechanisms that will help parties to move towards consensus and can be used as the basis for future progress. In the United States, an example of the former is the financial incentive that the Agency for Toxic Substances and Disease Registry is providing to investigators in the Great Lakes regions to coordinate their activities, particularly with respect to gathering common types of information, whether they are survey answers, disease registry data, or analytical techniques and interpretations. An example of the latter is the recommendation to the California Environmental Protection Agency to establish an ongoing group to coordinate approaches among the units within the agency.

At the international level, a number of efforts that are being coordinated through the International Programme on Chemical Safety provide examples of mechanisms that have shown promise in moving parties with often quite divergent viewpoints towards greater commonality. Some of these involve direct interactions among representatives of members of the European Community while others utilize the expertise of scientific organizations, such as the International Federation of Teratology Societies.

Participants at this meeting also emphasized that successful harmonization efforts, whether at the state, national, or international level, require increased attention to risk communication. This is needed to create an effective dialogue among the wide variety of stakeholders who may be involved in the harmonization process and also to maintain ongoing communication with members of the public who can have a large impact through their roles as consumers, constituents, and voters.

In sum the contributors to this book, representing state, national, and international organizations, have shown that there are not only many areas of agreement as to how risk assessment and risk management may be harmonized, but also that a wide variety of mechanisms exist to implement consensus recommendations. Future harmonization efforts will be needed to refine the precise nature of the frameworks that can be agreed upon and to identify those implementation mechanisms that will work the most effectively in the variety of risk situations that face the world.

Appendix A

Environmentally Sound Management of Toxic Chemicals, including Prevention of Illegal International Traffic in Toxic and Dangerous Products

UNITED NATIONS CONFERENCE ON ENVIRONMENT & DEVELOPMENT AGENDA 21, CHAPTER 19[1]

A1 INTRODUCTION

19.1. A substantial use of chemicals is essential to meet the social and economic goals of the world community and today's best practice demonstrates that they can be used widely in a cost-effective manner and with a high degree of safety. However, a great deal remains to be done to ensure the environmentally sound management of toxic chemicals, within the principles of sustainable development and improved quality of life for humankind. Two of the major problems, particularly in developing countries, are: (1) lack of sufficient scientific information for the assessment of risks entailed by the use of a great number of chemicals, and (2) lack of resources for assessment of chemicals for which data are at hand.

19.2. Gross chemical contamination, with grave damage to human health, genetic structures, and reproductive outcomes, and the environment, has in recent times been continuing within some of the world's most important industrial areas. Restoration will require major investment and development of new techniques. The long-range effects of pollution, extending even to the fundamental chemical and physical processes of the Earth's atmosphere and climate, are becoming understood only recently and the importance of those effects is becoming recognized only recently as well.

19.3. A considerable number of international bodies are involved in work on chemical safety. In many countries work programs for the promotion of chemical safety are in place. Such work has international implications, as chemical risks do not respect national boundaries. However, a significant strengthening of both national and international efforts is needed to achieve an environmentally sound management of chemicals.

19.4. Six program areas are proposed:

---

[1]Note: This is a final, advanced version of a chapter of Agenda 21, as adopted by the Plenary in Rio de Janeiro, on June 14, 1992.

The information contained within this appendix is in the public domain.

(a) expanding and accelerating international assessment of chemical risks;
(b) harmonization of classification and labeling of chemicals;
(c) information exchange on toxic chemicals and chemical risks;
(d) establishment of risk reduction programs;
(e) strengthening of national capabilities and capacities for management of chemicals; and
(f) prevention of illegal international traffic in toxic and dangerous products.

In addition, the short final subsection G deals with the enhancement of cooperation related to several program areas.

19.5. The six program areas are together dependent for their successful implementation on intensive international work and improved coordination of current international activities, as well as on the identification and application of technical, scientific, educational, and financial means, in particular for developing countries. To varying degrees, the program areas involve hazard assessment (based on the intrinsic properties of chemicals), risk assessment (including assessment of exposure), risk acceptability, and risk management.

19.6. Collaboration on chemical safety between the United Nations Environment Program (UNEP), the International Labor Organization (ILO), and the World Health Organization (WHO) in the International Programme on Chemical Safety (IPCS) should be the nucleus for international cooperation on environmentally sound management of toxic chemicals. All efforts should be made to strengthen this program. Cooperation with other programs, such as those of the Organization for Economic Cooperation and Development (OECD) and the European Communities (EC) and other regional and governmental chemical programs, should be promoted.

19.7. Increased coordination of United Nations bodies and other international organizations involved in chemicals assessment and management should be further promoted. Within the framework of IPCS, an intergovernmental meeting, convened by the Executive Director of UNEP, was held in London in December 1991 to explore this matter further.

19.8. The broadest possible awareness of chemical risks is a prerequisite for achieving chemical safety. The principle of the right of the community and of workers to know those risks should be recognized. However, the right to know the identity of hazardous ingredients should be balanced with industry's right to protect confidential business information. (Industry, as referred to in this chapter, shall be taken to include large industrial enterprises and transnational corporations as well as domestic industries.) The industry initiative on responsible care and product stewardship should be developed and promoted. Industry should apply adequate standards of operation in all countries in order not to damage human health and the environment.

19.9. There is international concern that part of the international movement of toxic and dangerous products is being carried out in contravention of existing national legislation and international instruments, to the detriment of the environment and public health of all countries, particularly developing countries.

19.10. In resolution 44/226 of 22 December 1989, the General Assembly requested each regional commission, within existing resources, to contribute to the prevention of the illegal traffic in toxic and dangerous products and wastes by monitoring and making regional assessments of that illegal traffic and its environmental and health implications. The assembly also requested the regional commissions to interact among themselves and to cooperate with the UNEP, with a view to maintaining efficient and coordinated monitoring and assessment of the illegal traffic in toxic and dangerous products and wastes.

A2   PROGRAM AREAS: SOME EXAMPLES

A2.1   (A) EXPANDING AND ACCELERATING INTERNATIONAL ASSESSMENT OF CHEMICAL RISKS

19.11. Assessing the risks to human health and the environmental hazards that a chemical may cause is a prerequisite to planning for its safe and beneficial use. Among the approximately 100 000 chemical substances in commerce and the thousands of substances of natural origin with which human beings come into contact, many appear as pollutants and contaminants in food, commercial products, and the various environmental media. Fortunately, exposure to most chemicals (some 1500 cover over 95% of total world production) is rather limited, as most are used in very small amounts. However, a serious problem is that even for a great number of chemicals characterized by high-volume production, crucial data for risk assessment are often lacking. Within the framework of the OECD chemicals program such data are now being generated for a number of chemicals.

19.12. Risk assessment is resource-intensive. It could be made cost-effective by strengthening international cooperation and better coordination, thereby making the best use of available resources and avoiding unnecessary duplication of effort. However, each nation should have a critical mass of technical staff with experience in toxicity testing and exposure analysis, which are two important components of risk assessment.

A2.1.1   Objectives

19.13. The objectives of this program area are:

1. To strengthen international risk assessment. Several hundred priority chemicals or groups of chemicals, including major pollutants and contaminants of global significance, should be assessed by the year 2000, using current selection and assessment criteria.
2. To produce guidelines for acceptable exposure for a greater number of

toxic chemicals, based on peer review and scientific consensus distinguishing between health- or environment-based exposure limits and those relating to socio-economic factors.

A2.1.2   Activities

Management-related activities
19.14. Governments, through the cooperation of relevant international organizations and industry, where appropriate, should:

1. Strengthen and expand programs on chemical risk assessment within the United Nations system IPCS (UNEP, ILO, WHO) and the Food and Agriculture Organization of the United Nations (FAO), together with other organizations, including the OECD, based on an agreed approach to data-quality assurance, application of assessment criteria, peer review, and linkages to risk management activities, taking into account the precautionary approach.
2. Promote mechanisms to increase collaboration among governments, industry, academia, and relevant nongovernmental organizations involved in the various aspects of risk assessment of chemicals and related processes, in particular the promoting and coordinating of research activities to improve understanding of the mechanisms of action of toxic chemicals.
3. Encourage the development of procedures for the exchange by countries of their assessment reports on chemicals with other countries for use in national chemical assessment programs.

Data and information
19.15. Governments, through the cooperation of relevant international organizations and industry, where appropriate, should:

1. Give high priority to hazard assessment of chemicals, that is, of their intrinsic properties as the appropriate basis for risk assessment.
2. Generate data necessary for assessment, building, *inter alia*, on programs of IPCS (UNEP, WHO, ILO), FAO, OECD, and EC and on established programs of other regions and governments. Industry should participate actively.

19.16. Industry should provide data for substances produced that are needed specifically for the assessment of potential risks to human health and the environment. Such data should be made available to relevant national competent authorities and international bodies and other interested parties involved in hazard and risk assessment, and to the greatest possible extent to the public also, taking into account legitimate claims of confidentiality.

International and regional cooperation and coordination
19.17. Governments, through the cooperation of relevant international organizations and industry, where appropriate, should:

1. Develop criteria for priority-setting for chemicals of global concern with respect to assessment.
2. Review strategies for exposure assessment and environmental monitoring to allow for the best use of available resources, to ensure compatibility of data, and to encourage coherent national and international strategies for that assessment.

A2.1.3   Means of implementation

Financial and cost evaluation
19.18. Most of the data and methods for chemical risk assessment are generated in the developed countries and an expansion and acceleration of the assessment work will call for a considerable increase in research and safety testing by industry and research institutions. The cost projections address the needs to strengthen the capacities of relevant United Nations bodies and are based on current experience in IPCS. It should be noted that there are considerable costs, often not possible to quantify, that are not included. These comprise costs to industry and governments of generating the safety data underlying the assessments and costs to governments of providing background documents and draft assessment statements to IPCS, the International Register of Potentially Toxic Chemicals (IRPTC), and OECD. They also include the cost of accelerated work in non-United Nations bodies, such as the OECD and EC.

19.19. The Conference Secretariat has estimated the average total annual cost (1993–2000) of implementing the activities of this program to be about $30 million from the international community on grant or concessional terms. These are indicative and order of magnitude estimates only and have not been reviewed by governments. Actual costs and financial terms, including any that are non-concessional, will depend upon, inter alia, the specific strategies and programs governments decide upon for implementation.

Scientific and technological means
19.20. Major research efforts should be launched in order to improve methods for assessment of chemicals as work towards a common framework for risk assessment and to improve procedures for using toxicological and epidemiological data to predict the effects of chemicals on human health and the environment, so as to enable decision makers to adopt adequate policies and measures to reduce risks posed by chemicals.

19.21. Activities include:

1. Strengthening research on safe/safer alternatives to toxic chemicals that pose an unreasonable and otherwise unmanageable risk to the environment or human health and to those that are toxic, persistent and bioaccumulative and that cannot be adequately controlled;

2. Promotion of research on, and validation of, methods constituting a replacement for those using test animals (thus reducing the use of animals for testing purposes);

3. Promotion of relevant epidemiological studies with a view to establishing a cause-and-effect relationship between exposure to chemicals and the occurrence of certain diseases; and

4. Promotion of ecotoxicological studies with the aim of assessing the risks of chemicals to the environment.

Human resource development

19.22. International organizations, with the participation of governments and nongovernmental organizations, should launch training and education projects involving women and children, who are at greatest risk, in order to enable countries, and particularly developing countries, to make maximum national use of international assessments of chemical risks.

Capacity-building

19.23. International organizations, building on past, present, and future assessment work, should support countries, particularly developing countries, in developing and strengthening risk assessment capabilities at national and regional levels to minimize, and as far as possible control and prevent, risk in the manufacturing and use of toxic and hazardous chemicals. Technical cooperation and financial support or other contributions should be given to activities aimed at expanding and accelerating the national and international assessment and control of chemical risks to enable the best choice of chemicals.

A2.2 (B) HARMONIZATION OF CLASSIFICATION AND LABELING OF CHEMICALS

A2.2.1 Basis for action

19.24. Adequate labeling of chemicals and the dissemination of safety data sheets such as ICSCs (International Chemical Safety Cards) and similarly written materials, based on assessed hazards to health and environment, are the simplest and most efficient way of indicating how to handle and use chemicals safely.

19.25. For the safe transport of dangerous goods, including chemicals, a comprehensive scheme elaborated within the United Nations system is in current use. This scheme mainly takes into account the acute hazards of chemicals.

19.26. Globally harmonized hazard classification and labeling systems are not yet available to promote the safe use of chemicals, *inter alia*, at the workplace or in the home. Classification of chemicals can be made for different purposes and is a particularly important tool in establishing labeling systems. There is a need to develop harmonized hazard classification and labeling systems, building on ongoing work.

A2.2.2 Objectives

19.27. A globally harmonized hazard classification and compatible labeling system, including material safety data sheets and easily understandable symbols, should be available, if feasible, by the year 2000.

A2.2.3 Activities

Management-related activities
19.28. Governments, through the cooperation of relevant international organizations and industry, where appropriate, should launch a project with a view to establishing and elaborating a harmonized classification and compatible labeling system for chemicals for use in all United Nations official languages, including adequate pictograms. Such a labeling system should not lead to the imposition of unjustified trade barriers.

The new system should draw on current systems to the greatest extent possible; it should be developed in steps and should address the subject of compatibility with labels of various applications.

Data and information
19.29. International bodies including, inter alia, IPCS (UNEP, ILO, WHO), FAO, the International Maritime Organization (IMO), the United Nations Committee of Experts on the Transport of Dangerous Goods, and OECD, in cooperation with regional and national authorities having existing classification and labeling and other information-dissemination systems, should establish a coordinating group to:

1. Evaluate and, if appropriate, undertake studies of existing hazard classification and information systems to establish general principles for a globally harmonized system;
2. Develop and implement a work plan for the establishment of a globally harmonized hazard classification system. The plan should include a description of the tasks to be completed, deadline for completion and assignment of tasks to the participants in the coordinating group;
3. Elaborate a harmonized hazard classification system;
4. Draft proposals for the standardization of hazard communication terminology and symbols in order to enhance risk management of chemicals and

facilitate both international trade and translation of information into the end-user's language; and

5.   Elaborate a harmonized labeling system.

A2.2.4   Means of implementation

Financial and cost evaluation

19.30. The Conference Secretariat has included the technical assistance costs related to this program in estimates provided in program area E. They estimate the average total annual cost (1993–2000) for strengthening international organizations to be about $3 million from the international community on grant or concessional terms. These are indicative and order of magnitude estimates only and have not been reviewed by governments. Actual costs and financial terms, including any that are non-concessional, will depend upon, inter alia, the specific strategies and programs governments decide upon for implementation.

Human resource development

19.31. Governments and institutions and nongovernmental organizations, with the collaboration of appropriate organizations and programs of the United Nations, should launch training courses and information campaigns to facilitate the understanding and use of a new harmonized classification and compatible labeling system for chemicals.

Capacity-building

19.32. In strengthening national capacities for management of chemicals, including development and implementation of, and adaptation to, new classification and labeling systems, the creation of trade barriers should be avoided and the limited capacities and resources of a large number of countries, particularly developing countries, for implementing such systems, should be taken into full account.

A2.3   (C) INFORMATION EXCHANGE ON TOXIC CHEMICALS AND CHEMICAL RISKS

A2.3.1   Basis for action

19.33. The following activities, related to information exchange on the benefits as well as the risks associated with the use of chemicals, are aimed at enhancing the sound management of toxic chemicals through the exchange of scientific, technical, economic, and legal information.

19.34. The London Guidelines for the Exchange of Information on Chemicals in International Trade are a set of guidelines adopted by governments with a view to increasing chemical safety through the exchange of information on chemicals. Special provisions have been included in the guidelines with regard to the exchange of information on banned and severely restricted chemicals.

19.35. The export to developing countries of chemicals that have been banned in producing countries or whose use has been severely restricted in some industrialized countries has been the subject of concern, as some importing countries lack the ability to ensure safe use, owing to inadequate infrastructure for controlling the importation, distribution, storage, formulation, and disposal of chemicals.

19.36. In order to address this issue, provisions for Prior Informed Consent (PIC) procedures were introduced in 1989 in the London Guidelines (UNEP) and in the International Code of Conduct on the Distribution and Use of Pesticides (FAO). In addition a joint FAO/UNEP program has been launched for the operation of the PIC procedures for chemicals, including the selection of chemicals to be included in the PIC procedure and preparation of PIC decision guidance documents. The ILO chemicals convention calls for communication between exporting and importing countries when hazardous chemicals have been prohibited for reasons of safety and health at work. Within the General Agreement on Tariffs and Trade (GATT) framework, negotiations have been pursued with a view to creating a binding instrument on products banned or severely restricted in the domestic market. Further, the GATT Council has agreed, as stated in its decision contained in C/M/251, to extend the mandate of the working group for a period of 3 months, to begin from the date of the group's next meeting, and has authorized the Chairman to hold consultations on timing with respect to convening this meeting.

19.37. Notwithstanding the importance of the PIC procedure, information exchange on all chemicals is necessary.

A2.3.2   Objectives

19.38. The objectives of this program area are:

1. To promote intensified exchange of information on chemical safety, use, and emissions among all involved parties.
2. To achieve by the year 2000, as feasible, full participation in and implementation of the PIC procedure, including possible mandatory applications through legally binding instruments contained in the Amended London Guidelines and in the FAO International Code of Conduct, taking into account the experience gained within the PIC procedure.

A2.3.3   Activities

Management-related activities

19.39. Governments and relevant international organizations with the cooperation of industry should:

1. Strengthen national institutions responsible for information exchange on toxic chemicals and promote the creation of national centers where these centers do not exist.
2. Strengthen international institutions and networks, such as IRPTC, responsible for information exchange on toxic chemicals.
3. Establish technical cooperation with, and provide information to, other countries, especially those with shortages of technical expertise, including training in the interpretation of relevant technical data, such as Environmental Health Criteria Documents, Health and Safety Guides and International Chemical Safety Cards (published by IPCS); monographs on the Evaluation of Carcinogenic Risks of Chemicals to Humans (published by the International Agency for Research on Cancer (IARC)); and decision guidance documents (provided through the FAO/UNEP joint program on PIC), as well as those submitted by industry and other sources.
4. Implement the PIC procedures as soon as possible and, in the light of experience gained, invite relevant international organizations, such as UNEP, GATT, FAO, WHO, and others, in their respective area of competence to consider working expeditiously towards the conclusion of legally binding instruments.

Data and information

19.40. Governments and relevant international organizations with the cooperation of industry should:

1. Assist in the creation of national chemical information systems in developing countries and improve access to existing international systems.
2. Improve databases and information systems on toxic chemicals, such as emission inventory programs, through provision of training in the use of those systems as well as software, hardware, and other facilities.
3. Provide knowledge and information on severely restricted or banned chemicals to importing countries to enable them to judge and take decisions on whether to import, and how to handle, those chemicals and establish joint responsibilities in trade of chemicals between importing and exporting countries.
4. Provide data necessary to assess risks to human health and the environment of possible alternatives to banned or severely restricted chemicals.

19.41. United Nations organizations should provide, as far as possible, all international information material on toxic chemicals in all United Nations official languages.

International and regional cooperation and coordination

19.42. Governments and relevant international organizations with the cooperation of industry should cooperate in establishing, strengthening, and expanding, as appropriate, the network of designated national authorities for the exchange of information on chemicals and establish a technical exchange program to produce a core of trained personnel within each participating country.

A2.3.4　Means of implementation

Financing and cost evaluation
19.43. The Conference Secretariat has estimated the average total annual cost (1993–2000) of implementing the activities of this program to be about $10 million from the international community on grant or concessional terms. These are indicative and order of magnitude estimates only and have not been reviewed by governments. Actual costs and financial terms, including any that are non-concessional, will depend upon, inter alia, the specific strategies and programs governments decide upon for implementation.

Appendix B

Risk Assessment and Risk Management in Regulatory Decision Making: Commission on Risk Assessment and Risk Management Draft Report, June 13, 1996

COMMISSION ON RISK ASSESSMENT AND RISK MANAGEMENT

Gilbert S. Omenn (Chairman), Dean, School of Public Health and Community Medicine, University of Washington, Seattle, WA

Alan C. Kessler (Vice-Chairman), Partner, Buchanan Ingersoll, Philadelphia, PA

Norman T. Anderson, Director of Research, American Lung Association of Maine, Augusta, ME

Peter Y. Chiu, Senior Physician, Kaiser Permanente, Milpitas, CA

John Doull, Professor, Department of Pharmacology, Toxicology and Therapeutics, Kansas University Medical Center, Kansas City, KS

Bernard Goldstein, Director, Environmental and Occupational Health Sciences Institute and Chairman, Department of Environmental and Community Medicine, UMDNJ-Robert Wood Johnson Medical School, Piscataway, NJ

Joshua Lederberg, President Emeritus, Rockefeller University, New York, NY

Sheila McGuire, President, Iowa Health Research Institute, Boone, IA

David Rall, Former Director, National Institute of Environmental Health Sciences, Washington, DC

Virginia V. Weldon, Senior Vice President for Public Policy, Monsanto Company, St. Louis, MO

Staff

Gail Charnley, Executive Director

Sharon Newsome, Associate Director

Joanna Foellmer, Program Specialist and Designated Federal Official

B1 EXECUTIVE SUMMARY

Public opinion polls have consistently shown strong support throughout the United States for effective environmental stewardship and for identifying and addressing risks to the environment, public health, and worker health. At the same time, many citizens and local officials are demanding greater attention to

---

This appendix is a US Government work and, as such, is in the public domain in the United States of America.

priorities and costs. There is an emerging national vision of sustainable development for our environment, our economy, and our society, which this Commission shares. Regulatory agencies, businesses, environmental and public health advocates, and communities deserve credit for well documented gains in air quality, water quality, habitat protection, product safety, waste disposal, recycling, and pollution prevention achieved over the last 25 years. The Commission values and seeks to sustain such gains. Our findings and recommendations reflect an increasing need to recognize and capitalize on lessons learned and our intent to stimulate an even more efficient, more effective, risk-based means of protecting public health and the environment.

The Commission on Risk Assessment and Risk Management was mandated by Congress in the Clean Air Act Amendments of 1990 'to make a full investigation of the policy implications and appropriate uses of risk assessment and risk management in regulatory programs under various Federal laws to prevent cancer and other chronic human health effects which may result from exposure to hazardous substances'. The Commission began meeting in May 1994 and held hearings across the United States, obtaining information and insights that made important contributions to our deliberations and to our findings and recommendations. With this draft report, we introduce a framework for making risk management decisions; we evaluate and make recommendations about the uses and limitations of risk assessment, economic analysis, risk management, and regulatory decision making; and we address selected activities of specific regulatory agencies and programs. The Commission continues to seek comments as we refine our recommendations for the final report to Congress and the President of the United States.

B2 A NEW RISK MANAGEMENT FRAMEWORK

The Commission has adopted a unique risk management perspective to guide investments of precious public sector and private sector resources in risk-related research, risk assessment, risk characterization, and risk reduction. We recognize that it is time to modify the traditional approaches to assessing and reducing risks that have relied on a chemical-by-chemical, medium-by-medium, risk-by-risk strategy. While risk assessment has been growing more complex and sophisticated, the output of risk assessment for the regulatory process often seems too focused on refining assumption-laden mathematical estimates of small risks associated with exposure to individual chemicals rather than on the overall goal–risk reduction and improved health status. Scientists, federal agencies, the National Academy of Sciences/National Research Council, and many other organizations have issued many reports with recommendations for improving health risk assessment. Despite many years of managing risks, however, there have been few systematic attempts to examine the role of risk assessment itself in risk management and health and environmental protection.

No generally accepted framework or principles for making risk management decisions has emerged.

We propose a systematic, comprehensive framework that can address various contaminants, media, and sources of exposure, as well as public values, perceptions, and ethics, and that keeps the focus on the risk management goal. The new risk management framework comprises six stages (see Figure B1):

- formulate the problem in broad context
- analyze the risks
- define the options
- make sound decisions
- take actions to implement the decisions
- perform an evaluation of the effectiveness of the actions taken.

The framework explicitly embraces collaborative involvement of stakeholders; the process can be refined and its conclusions can be changed as important new information is acquired.

The framework requires first that a potential or current problem be put into

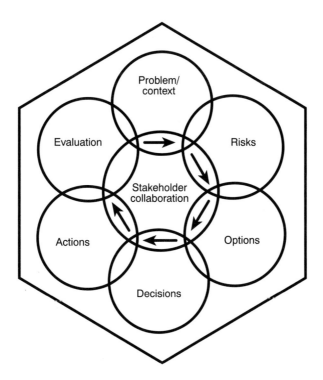

Figure B1   The Commission's risk management framework.

a broader context of public health or environmental health and that the inter-dependence of related multimedia problems be identified. For example, the risks associated with the hazardous air pollutants regulated at one industrial facility or category of facilities can be considered in context with the risks associated with other stationary and mobile sources that emit the same pollu-tants. The next layer of context would be provided by comparisons with risks associated with other important air pollutants, such as particles and carbon monoxide, emitted by the same sources. A multimedia context would lead to a comprehensive plan that includes risks associated with air, water, and solid waste in a particular geographic area.

The framework actively engages stakeholders, especially at the initial stage of formulating the problem; we want to go beyond worker and community right to know requirements and make stakeholders partners in risk assessment and risk management. In later stages of the framework, risk managers and stake-holders investigate the risks, including cumulative risks to human and environ-mental health; risk reduction options are identified, and potential consequences evaluated, including the benefits, costs, and social, cultural, ethical, political, and legal dimensions of each option; and the responsible agency then makes a decision that reflects input from stakeholders, implements a risk reduction action, and seeks credible evaluation of the outcome. As new information or new technology becomes available, the problem can be redefined, and the risk management process repeated, if appropriate.

This framework can help to improve the cumbersome, fragmented risk management approach often used by the federal regulatory agencies—an approach that resulted from the patchwork of Congressional statutes that have been passed over the last 25 years to address individual risks. Coordination within and among agencies and among Congressional committees and subcom-mittees can advance the more comprehensive proposed framework without a new, overarching environmental statute. The framework is also applicable to risk management activities carried out by public and private entities at the state, regional, and local levels. Despite potential obstacles, we believe that imple-mentation of this framework will enable the country to manage risks more effec-tively and more efficiently and to make progress toward the goal of sustainable development.

B3 USES AND LIMITATIONS OF RISK ASSESSMENT

The Commission considers risk assessment a useful analytic process that provides valuable contributions to risk management, public-health, and environ-mental-policy decisions. Risk assessment was developed because Congress, regulators, and the public require scientists to go beyond scientific observations of the relationships between exposures to chemicals and pollutants and their effects on people, the environment, or test systems, and to rely on many scientific inferences and assumptions to answer social questions about what is unsafe. When basic judgements regarding a chemical's toxicity to humans are unresolved, however, sophisticated and complex risk assessments might not be immediately helpful to risk managers. We recommend that the performance of risk assessments be guided by an understanding of the issues that will be important to managers' decisions and to the public's understanding of what is needed to protect public health and the environment.

B3.1 USE OF GOOD SCIENCE IN TOXICITY ASSESSMENTS

The Commission recognizes that important advances are being made in the scientific basis for risk assessment. Further developments will improve the recognition and estimation of risks to humans associated with chemical and other exposures in the environment and provide biologic markers for measuring exposure, early effects, and variation in susceptibility. We recommend the use of all relevant peer-reviewed information about a chemical's mode of action in evaluating the weight of the scientific evidence supporting its toxicity in humans. We support current agency efforts to distinguish more clearly between experimental findings in rodent or other bioassays that are predictive for humans and findings that are not. We recognize that risks from microbial and radiation exposures, not just chemical exposures, need to be addressed, and we recommend the evaluation of a common metric to assist comparative risk assessment, risk communication, and risk characterization related to both carcinogens and noncarcinogens.

B3.2 USE OF REALISTIC SCENARIOS IN EXPOSURE ASSESSMENTS

The Commission supports basing risk management decisions on exposure assessments derived from realistic scenarios. Agencies should continue to move away from using the hypothetical 'maximally exposed individual' to evaluate whether a risk exists, toward more realistic assumptions based on available scientific data, as they have done in recent analyses. We recommend use of analytic methods that, when data permit, combine the many characteristics of probable exposure into an assessment of the overall population's exposures. Where possible, exposure assessments should include information about specific groups, such as infants, children, pregnant women, low-income groups, and minority-group communities with exposures tied to particular cultural or social practices. Stakeholders can provide information about patterns and sources of exposure that otherwise might be neglected.

B3.3 RECOGNITION OF RISK ASSOCIATED WITH CHEMICAL MIXTURES

We agree with testimony that we need data and risk estimates about chemical mixtures and combined chemical–microbial–radiation exposures, because people are exposed to multiple hazards. We recommend direct toxicity assays of environmental mixtures.

B4 USES AND LIMITATIONS OF ECONOMIC ANALYSIS

The Commission supports the use of economic analysis as a consideration, but not as an overriding determinant of risk management decisions. Both human health and ecological benefits should be accounted for when the consequences of actions to reduce emissions, exposures, and risks are being evaluated. We call for explicit descriptions of the assumptions, data sources, sources of uncertainty, and distributions of benefits and costs across society associated with economic analyses, in parallel with the descriptions associated with risk assessments.

B5 RISK MANAGEMENT AND REGULATORY DECISION MAKING

Risk assessment and economic analysis provide only part of the information that risk managers use—with information about public values and statutory requirements—to make decisions about the need for and methods of risk reduction. The wide array of statutes and their implementing regulations have resulted in different definitions of negligible and unacceptable risk, and the use of risk assessment has differed in decision making and regulatory programs.

B5.1 IMPROVEMENT OF RISK COMMUNICATION

In communicating to various audiences about risks, risk assessors must seek a two-way interaction, learning about patterns of exposure, gaining an understanding of the different perceptions people have of what is a negligible risk and what is an unacceptable risk, and describing risks and uncertainties openly and understandably. Relying on overprecise single estimates of risk is unwarranted.

We support the use of comparisons of specific risks related to a proposed action with emphasis on chemically related agents, different agents to which humans might be exposed in similar ways, different sources of exposure to the same agents, and different agents that produce similar effects. Such context can help all stakeholders, including risk assessors, to understand the potential benefit of reducing exposures to an agent. We recommend that such risks be expressed in terms of potential adverse effects per year in a given community or exposed population, as well as per hypothetical lifetime.

B5.2 BRIGHT LINES

Bright lines are specific exposure concentrations or levels of risk that are meant to provide a clear distinction between what is considered safe and what is not. Bright lines can be useful as guideposts or goals for decision making, but should not be applied inflexibly because of uncertainty about risks and susceptibility. We support the use of sets of bright lines to protect both the general population and specific populations potentially at higher risk, such as children and pregnant women. We do not support efforts by Congress to legislate particular bright lines.

B5.3 PEER REVIEW

We support the efficient use of peer review, with care to exclude conflicts of financial interest, for both risk assessment and economic analysis. Peer-review quality and effectiveness should be evaluated regularly. We urge Congress to match resources to its demands on agencies for research, risk assessment, and economic analysis and to allow the agencies considerable discretion in allocating resources to their peer-review efforts.

B5.4 STANDARDS OF JUDICIAL REVIEW

We recommend that judicial review be limited, as now, to final agency action, and that the existing arbitrary-and-capricious standard be retained as is.

B6 RECOMMENDATIONS FOR AGENCIES

The Commission developed findings and recommendations about several federal agencies and programs, partly to illustrate our general recommendations, partly to address inconsistencies, and partly to try to assist Congress and the agencies on particular matters.

B6.1 ENVIRONMENTAL PROTECTION AGENCY

In the 1990 amendments to the Clean Air Act, Congress mandated that this Commission review and make recommendations on the analysis and treatment of residual risks associated with section 112 hazardous air pollutants after the completion of the current technology-based risk reduction program. We present a tiered approach to set priorities for this huge effort. We recommend that residual risks associated with hazardous air pollutants be considered in the context of risks associated with the same pollutants from other sources and in the context of other risks to health.

We recommend more frequent determinations of future land use at the start of Superfund-site risk assessments, and updating of the Toxic Substances Control Act to reflect advances in the understanding of chemical toxicology. We endorse a comprehensive watershed-management approach to managing risks under the Clean Water Act.

B6.2 FOOD AND DRUG ADMINISTRATION

We propose a substantial modification of the 'Delaney clause', international harmonization of risk assessment and clinical-trial protocols for pharmaceuticals, and restoration of the FDA's authority to require scientific evidence supporting health claims for dietary supplements.

B6.3 DEPARTMENT OF AGRICULTURE

We recommend that risk assessment and benefit–cost analysis be performed early in the rule-making process instead of at the decision stage.

B6.4 DEPARTMENT OF ENERGY AND DEPARTMENT OF DEFENSE

We propose further development and evaluation of risk-based approaches to priority setting and budget making.

B7 INTRODUCTION

The Commission on Risk Assessment and Risk Management was mandated by Congress in the 1990 amendments to the Clean Air Act to address risks that are regulated under the many laws aimed at protecting the environment and protecting the health and safety of the American people from potentially dangerous exposures to chemicals and other hazardous substances and objects in air, water, food, the workplace, and consumer products. Of the 10 members of the Commission, three were appointed by the president, six by the majority and minority leaders of the House and Senate, and one by the president of the National Academy of Sciences. The Commission's mandate is summarized in the following phrases:

- assess uses and limitations of risk assessment
- evaluate exposure scenarios for risk characterization
- determine how to describe and explain uncertainties
- enhance strategies for risk based management decisions
- review desirability of consistency across federal programs

The Commission was also asked to comment on the conclusions of *Science and Judgment in Risk Assessment* (National Research Council 1994) and to make recommendations about peer review.

Congress decided to create the Commission when agreement could not be reached, during drafting of the 1990 Clean Air Act Amendments, on the best way for the US Environmental Protection Agency (EPA) to determine whether any significant risks to human health will remain after technology-based controls are implemented to reduce hazardous-pollutant emissions from stationary sources and, if so, what to do about those residual risks. There was disagreement about the risk assessment techniques and assumptions that should be used to estimate residual risks, about the benchmarks that should be used to distinguish between negligible and unacceptable risks, and about the risk management methods that should be used to mitigate unacceptable risks. But the Commission's mandate was not restricted to evaluating air pollution, the particulars of the Clean Air Act, or the EPA. Rather, it was limited to 'cancer and other chronic human health effects', so we did not address environmental problems, such as global climate change, ozone depletion in the stratosphere, or protection of wetlands and other habitats. We do note, however, that human health depends on a healthy environment, that the general approaches of health risk assessment are applicable to ecological risk assessment, and that benefit–cost analyses should assess all benefits, not just human health benefits.

B8 VISION

Through its deliberations, the Commission developed a shared vision of sustainable goals for our environment, our economy, and our society. Like the National Commission on the Environment (1992) and the President's Council on Sustainable Development (1996), we seek a convergence of economic and environmental goals and actions. We recognize that special sensitivity is required to encompass the diverse socioeconomic status and cultural practices of United States. We seek a comprehensive, risk based approach that puts specific actions in a public health and ecological context.

B9 BACKGROUND

As a result of public recognition of environmental problems and translation of that recognition into effective action, tremendous progress has been achieved during the last 25 years in improving air quality, water quality, safety at work, safety of consumer products (including drugs and foods), testing of new chemicals before they are introduced into commerce, cleanup and disposal of hazardous wastes, and scientific study of health effects and ecological effects of chemicals, radiation, and microorganisms. Improvements in public health historically have come primarily from environmental interventions, such as proper waste disposal and hygiene, quarantines, clean water, and vaccines. Although many federal environmental laws share a primary goal of protecting the public's health and the environment, most environmental statutes have been media specific and have relied on regulatory approaches rather than public health approaches.

We know that the gains of the last 25 years can be sustained only by continued action, especially as the economy and the population grow and new technologies are introduced, and we believe that the effort to sustain them will be most effective if regulatory and public health agencies work together.

B10   RISK ASSESSMENT

Risk is a combination of the probability of an event—usually an adverse event—and the nature and severity of the event. We deal with risks all the time in everyday life—risks to our health, our environment, our pocketbooks, our social relationships. Risk is time related, from immediate consequences of various actions or lack of action to consequences over a lifetime for an individual and much longer periods for the whole society or the planet. We make decisions to avoid risks, to reduce risks, to reduce the consequences of events, and to insure against the financial consequences of risks. We tend to downplay

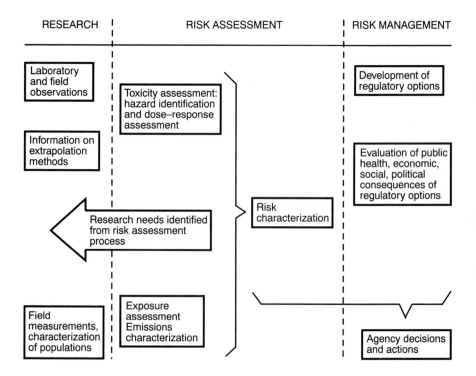

Figure B2   Elements of risk assessment and risk management. Source: *Science and Judgment in Risk Assessment* (National Research Council 1994). Reprinted with permission.

some risks; we find others frightening. Of course, people vary in those assessments, and their actions or concerns tend to vary accordingly. Often, the people who face specific risks are different from the people who benefit from the events involved in the risks, leading to conflict and litigation over proposed actions. Risk assessment itself has become controversial because of its important role in the protection of human health and the environment.

A generally accepted framework and nomenclature for health risk assessment was established in 1983 by a National Academy of Sciences committee report, *Risk Assessment in the Federal Government: Managing the Process* (National Research Council 1983). The now universally recognized four-step framework for characterizing the likelihood of adverse health effects from particular chemical exposures is described briefly below and shown in the context of scientific issues and regulatory impact in Figure B2.

- *hazard identification*: determine the identities and quantities of environmental contaminants present that can pose a hazard to human health
- *dose–response assessment*: evaluate the relationship between contaminant exposure concentrations and the incidence of adverse effects in humans
- *exposure assessment*: determine the conditions under which people could be exposed to contaminants and the doses that could occur as a result of exposure
- *risk characterization*: describe the nature of adverse effects that can be attributed to contaminants, estimate their likelihood in exposed populations, and evaluate the strength of the evidence and the uncertainty associated with them

The Commission was directed to focus on what Congress called 'chronic health effects', meaning effects that do not occur immediately—such as injuries from falling off a construction platform—but that are the cumulative result of repeated exposures that might take months, years, or decades to become manifest as health problems. Risks from chronic exposures arise from activities associated with the use and production of food, energy, industrial and consumer goods, and from the wastes produced through daily living. We recognize that voluntary uses of specific consumer products are also major contributors to death and poor health. Cigarette-smoking leads the list by a wide margin, accounting for an estimated 400 000 deaths every year (McGinnis and Foege 1993). Use of alcoholic beverages accounts for about 100 000 deaths, and motor-vehicle collisions for about 25 000 deaths. As many as 60 000 deaths per year are estimated to be attributable to airborne fine particles. Many activities individually contribute little to overall public health risks but substantially when viewed collectively. For example, 60 000 deaths per year have been attributed to occupational and environmental chemical exposures of all types.

Of all the causes of death, the most salient for most people is cancer; it is important to recognize that cancer has multiple causes and is not a single disease. However, cancer is not the only cause of health concerns associated with environmental pollutants. Reproductive impairments, birth abnormalities, asthma and other forms of airway hyperactivity, and effects on all the organ systems of the body warrant serious attention from a risk management perspective. Even if those health effects have modest impacts on mortality, they are important determinants of our quality of life.

Risk assessment goes beyond scientific observations of exposures and effects in people, animals, or test systems to try to answer social questions about what is unsafe. There is a difference between what can be studied experimentally or be observed directly and what represents policy-driven extrapolation based on scientific inferences and many assumptions. The 1994 National Research Council report *Science and Judgment in Risk Assessment* captured this combination of science and values in its title.

B11 RISK MANAGEMENT

We face a huge challenge to manage comprehensively the health risks associated with the vast array of pollution-generating activities in the United States. Our regulatory agencies are expected to control, down to an extremely low level, the potential cancer risks, for example, associated with each of those individual activities; a limit of less than one extra cancer death from a particular chemical per million persons exposed over a 70-year lifetime is generally used for screening purposes and when exceeded, might serve as a justification for seeking exposure monitoring data to characterize risks more accurately. Risk criteria used in regulating occupational exposure to specific chemicals often correspond to about one extra cancer death out of every 1000 workers exposed over a working lifetime. For noncancer risks, regulatory agencies aim to reduce exposures to below presumed thresholds for adverse effects.

As directed by Congress and reinforced by the Clinton Administration, we have framed our analyses and recommendations from the perspective of risk management. How do we use the tools of risk assessment and of economic analysis and consider social and cultural information to make better, more efficient, more understandable, and less costly decisions about reducing risks that are judged to be too high? How do we compare risks and risk reduction actions of various kinds to determine which deserve higher priority? What are the community, public health, and environmental contexts for formulating a particular problem, characterizing its risks, deciding what to do about it, and evaluating the impact of actions taken? It is crucial to reach out to affected parties and communities to obtain knowledge about the nature of past and present exposures and to understand their concerns and perceptions about the risks under discussion and related risks. Communication about risks is a two-way process.

To address those questions, the Commission proposes a comprehensive risk management framework for making decisions about reducing risks to public health and the environment. The process includes detailed consideration of risk and cost and provides a context for social and cultural considerations. One salient feature of the framework is its explicit involvement of stakeholders in decisions about how to reduce the risks that affect them—through consensus or despite disagreement—depending on the circumstances. Another salient feature is the integrated, multimedia approach the framework takes to address multiple risks instead of individual risks.

B12 OUR REPORT

This report is the product of the Commission's deliberations and evaluations since May 1994 and constitutes a response to concerns of those who provided testimony before the Commission and issue papers prepared for the Commission by several experts.

B13 FRAMEWORK FOR RISK MANAGEMENT

It is time to change the traditional approaches to assessing and reducing environmental, health, and safety risks, which have relied on a chemical-by-chemical, medium-by-medium, risk-by-risk strategy. That strategy evolved from multiple, unrelated statutory requirements from the various Congressional subcommittees that have jurisdiction over agencies responsible for protecting health and the environment. The result is our highly fragmented and adversarial system of conflicting actions that ignores the interdependence of environmental components, emphasizes the differences instead of the similarities between cancer and other health effects, and investigates risks associated with individual purified chemicals instead of environmental mixtures of chemicals.

Many effective risk management decisions certainly have been made, but many other decisions have left stakeholders unhappy or problems only partly addressed. Testimony received from the US Department of Energy (DOE) Office of Integrated Risk Management, the EPA deputy and assistant administrators, and the public repeatedly emphasized the need to address multiple chemical exposures and called for our environment to be addressed as a system, not as a fragmented collection of individual risks. A related difficulty is the need to combine characterizations of risks to health and the environment with values, perceptions, and concerns of affected parties, especially nontechnical people and communities.

Moving toward integrated, effective environmental management requires a risk management framework that can engage a wide range of stakeholders and address the interdependence and cumulative effects of various problems. The framework must have the capacity to address various media, contaminants, and sources of exposure and an array of public values, perceptions, and ethics. It should be sufficiently understandable to be adopted and used by risk managers in a wide variety of situations and lead to acceptable and effective decisions. It should be flexible so that its use can be matched to the importance of the decisions to be made. Full implementation of the framework for federal programs will lead to a need for Congressional authorization and funding; however, much progress can be made with existing statutes.

The overarching goal of the Commission's framework for integrated risk management is to move beyond 'end-of-the-pipe' command-and-control approaches to environmental protection toward means of achieving sustainable development. Such an ethic of environmental stewardship requires an understanding of the connections between environmental health, human and economic well-being, and the processes by which our society's actions create long-term changes, both beneficial and adverse.

The Commission believes that the integrated risk management framework described here is timely. It is consistent with regulatory reform goals of stimulating economic progress while improving the effectiveness of environmental protection and sustaining the accomplishments of the last 25 years of environmental regulation. Thus, it is consistent with the goals of the President's Council on Sustainable Development—to ensure that every person enjoys the benefits of clean air, clean water, and a healthy environment while maintaining economic prosperity and creating full opportunity for citizens, businesses, and communities to participate in and influence the natural resource, environmental, and economic decisions that affect them (President's Council on Sustainable Development 1996). Its insistence on collaborative stakeholder involvement and empowerment, its commitment to place each problem in a public health context, and its use, when appropriate, in an iterative manner to refine regulatory decisions make it applicable to diverse environmental regulatory problems. Although the Commission's mandate was to evaluate risk management decision making at the federal level, the framework is applicable to all levels of decision making.

The framework reflects the importance of 'participatory democracy' in resolving environmental dilemmas (Ruckelshaus 1995). It is consistent with testimony calling for risk management partnerships among government, industry, and the public. For example, Walter Fields, of the National Association for the Advancement of Colored People (NAACP) in New Jersey, urged the Commission to define specific steps needed to bring communities into the risk assessment and risk management processes and to enable communities to engage in honest dialogues with industries. The Commission received similar testimony from a variety of people all over the United States, including: Michael McCloskey, chairman of the Sierra Club; Linda Greer, of the Natural Resources Defense Council; Mark Van Putten, of the National Wildlife Federation; Peter Raven, of the Missouri Botanical Garden; Ronald Selph, mayor of Granite City, Illinois; Carol Henry, director of the DOE Office of Risk Management; and Phillip Lewis, of Rohm and Haas Company, Philadelphia. The framework incorporates various principles and recommendations set forth by the Carnegie Commission report *Risk and the Environment: Improving Regulatory Decision Making* (1993), the National Commission on the Environment report *Choosing a Sustainable Future*, the National Research Council report *Science and Judgment in Risk Assessment*, the Annapolis Accords for Risk Analysis, the Harvard Group on Risk Management Reform, and the National Academy of Public Administration report *Setting Priorities, Getting Results* (1995).

B13.1 FINDING

After many years of management of environmental, health, and safety risks in the United States, there is still no generally accepted or uniformly applied framework or set of principles for making risk management decisions. Current efforts to manage risks are often fragmented and sometimes in conflict, often

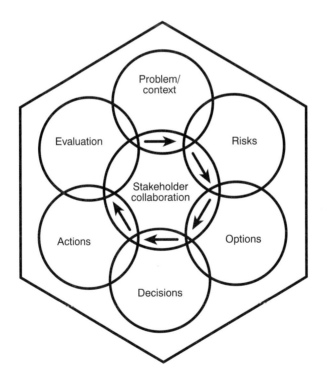

Figure B3 Framework for risk management.

reflecting different statutes. In addition, there is no systematic process for integrating public values, perceptions, ethics, and other cultural considerations into risk management decisions.

B13.2 RECOMMENDATION

A systematic, comprehensive risk management framework should be used to reduce environmental, health, and safety risks. The Commission's framework comprises six stages (Figure B3): (1) formulating the problem in broad context; (2) analyzing the risks; (3) defining the options; (4) making sound decisions; (5) taking actions to implement the decisions; and (6) evaluating the effects of the actions taken. The framework can be used iteratively and must embrace collaborative involvement of stakeholders.

B14 RATIONALE

B14.1 THE FRAMEWORK

The Commission's risk management framework constitutes a comprehensive strategy for reducing risks to public health, safety, and the environment. Each stage involves a different set of questions. The following is a description of how the six stages would operate and the collaborative and iterative processes that would occur throughout them.

B14.1.1 Problem/context

What is the problem? What is its context? Who is responsible for managing the problem, and who are the stakeholders?

A potential or existing problem might be identified on the basis of: environmental monitoring; emissions inventories; disease surveillance and epidemiologic observation; unexplained illnesses; a permit application; a bad odor; a need for national standards to control contaminant concentrations in air, water, soil, or food; or some other public concern. Attention might be focused on a 'symptom' of an underlying problem.

The problem would be examined not just in a medium- and pollutant-specific manner, but in a comprehensive, multimedia, public health context. Potential relationships among different problems are identified and considered. For example, the degradation of an aquatic ecosystem can be caused not only by point sources of water pollution but also by nonpoint sources, such as urban and agricultural runoff. It can also be affected by land-disturbance activities, including logging and grazing, construction of dams and reservoirs, diversion of surface-water and groundwater flows for domestic and agricultural uses, overfishing, and introduction of exotic species. Deposition of air pollutants, such as nitrogen, lead, cadmium, chromium, mercury, and radionuclides, can also contribute to the problem.

Stakeholders are identified at the problem/context stage and are relied on heavily for problem identification and characterization. Risk managers can be people or institutions at the federal, state, or community level, depending on the problem's context. Once there is collective appreciation of (and possibly consensus on) the characterization of the problem, risk management goals and objectives can be defined and pursued.

Appropriate contexts for a problem are likely to be situational. In some cases, the context can be overall public health. In other cases, it might be other risks. In still others, it might be its relationship to the interdependence of different problems (such as the degradation of an aquatic ecosystem, as described above).

An example of formulating a problem in its context might start with the consideration of the risks associated with hazardous air pollutants regulated at one industrial facility or category of facilities in the context of risks associated with stationary and mobile sources that emit the same pollutants in the same geographic area. The next layer of context would be comparisons with risks associated with other important air pollutants, such as particles and carbon monoxide. A multimedia context would lead to a comprehensive plan that includes risks associated with air, water, and solid waste in the region.

Putting a problem in its context will be needed when residual risks associated with hazardous air pollutants are characterized after technology-based controls are implemented under the Clean Air Act. For example, the EPA has promulgated a maximum-available-control-technology (MACT) standard for the petroleum refinery industry. That standard was promulgated partly on the basis of the EPA's finding that emissions from petroleum refineries potentially pose a leukemia risk to exposed populations because they contain benzene, a known human carcinogen. The standard will reduce the emissions of benzene and other hazardous air pollutants emitted by the source category. After the standard is implemented, residual-risk assessments will include a calculation to determine whether a leukemia risk that can be attributed to benzene remains. In addition to being emitted from petroleum refineries, benzene is emitted as exhaust from motor vehicles. In fact, emissions from mobile sources represent the largest single source of airborne benzene in the United States. Residual risks associated with the benzene emitted from a particular petroleum refinery could be compared with those associated with benzene from mobile sources and any other important sources that might exist in the geographic area of interest, including consumer products used in the home. The advantage of considering the risk in that context is that if the residual leukemia risk contributed by the petroleum refinery proves to be significant in comparison with the leukemia risk contributed by other sources, risk reduction efforts can be focused on modifying the refinery further. If it proves insignificant in comparison, more effective and more efficient means of risk reduction that focus on the larger contributors of risk might be possible.

B14.1.2 Risks

What risks does the problem pose to public health?

Risk would be determined by considering the nature, likelihood, and severity of adverse effects on human health, the environment, or public welfare, such as economic well-being or aesthetics. Risks would be evaluated primarily by scientists and risk managers with input from stakeholders. Community stakeholders should be consulted at this stage to help to identify groups with high exposures so that appropriate exposure assessments can be designed. The factual and scientific basis of the problem would be articulated and incorporated, along with subjective perceptions of the problem, into a characterization of the risks to human and environmental health and consideration of cultural and societal values, quality of life, and environmental equity. Health and ecological risks would be treated both qualitatively and quantitatively. The nature of the adverse effects, their severity, their reversibility or preventability, and the possibility of multiple effects must be understood before complex estimates of the magnitude of the risks and their uncertainties are presented. Cumulative risks associated with related problems would be identified, where possible. Indirect effects on human health through disruption of the environment also would be considered.

B14.1.3 Options

What can and should be done about the problem? What are the potential consequences and expected benefits of intervention? Are there other ways to reduce similar health effects in the same population or similar ecological effects in the same region? What are the estimated costs of each option?

Approaches to addressing the problem would be identified by stakeholders, regulators, and scientists. A variety of regulatory and nonregulatory alternatives would be considered, such as permits, enforcement actions, pollution prevention, recycling, market incentives, voluntary reductions, and education. Institutional, financial, and other arrangements for implementing each approach would be identified. The extent of risk reduction expected and the relationship between the costs and benefits of each approach would be determined and compared. Cultural, ethical, political, and legal dimensions would be considered. Potential impacts of each approach would be characterized, including possible adverse effects on workers, the community, or the environment.

B14.1.4 Decision

What is the best solution to the problem? How can that decision or set of decisions be reached? Who should make that decision? Will the actions required be compatible with the Unfunded Mandates Act of 1995?

The most feasible, effective, acceptable, and cost-effective approaches to mitigating the problem would be identified, with the participation of affected and responsible parties. A mechanism for conflict resolution, or for reaching closure in the absence of consensus, might be needed. It is important to acknowledge that this framework will not always result in a consensus among those involved in the process. In fact, participation, negotiation, and attempted compromise sometimes can result in a hardening of opposite positions, a breakdown in negotiations, frustration with the process, and an inability to reach agreement. Those difficulties in reaching a decision should be viewed not as a failure of the process envisioned by the framework, but simply as a recognition that in some instances, notwithstanding the best efforts of all affected parties, consensus will not be achievable. At some point, the responsible regulatory authority must make its decision, including a decision to defer, if opposition is too strong or too credible. Deferral would require a later decision of whether to repeat the process from the beginning or to go on to other pressing needs.

B14.1.5 Actions

How can the decision be implemented rapidly and flexibly?

The action that has been chosen to address the problem is explained and taken. Several actions might be needed in various circumstances. Actions might be taken by public agencies, businesses, industries, and private citizens, alone or in combination. Objections or reassessments, even at this stage, may trigger an iteration of the process.

B14.1.6 Evaluation

How effective are the actions?

Too often, actions are mandated but there is little follow up to ensure that they are taken, to analyze effectiveness and cost, or to compare the findings with estimates made in the decision-making stage. The effect of a chosen action on the problem can be characterized through monitoring and surveillance, through discussions with stakeholders, and through analyses of relationships between interventions and trends in health or environmental indicators. Criteria should be specified in advance. On the basis of evaluation, the original problem can be redefined, the actions reconsidered, and the various stages repeated if appropriate.

If the evaluation of the impact of the actions on a problem finds it unsatisfactory, another iteration of the process might be needed. But few effects on risk lend themselves easily to measurement and validation. To some extent, monitoring and surveillance can enable the study of relationships between action and effect, but often such relationships are detectable only when the margin between actual exposures and exposures associated with the health or ecological effect of concern is narrow or the health effect of interest is particularly rare. Most public health risks are already low, compared with such measurable effects as occupational injuries or motor-vehicle collisions. For example, suppose an action lowers the lifetime incremental risk of developing cancer because of a particular exposure from 1 in 10 000 to 1 in 1 000 000. No health study or surveillance activity could be designed to measure the effectiveness of an action with such a small change, because cancer would be the cause of death in 24% of the population in either event. Conclusions about effectiveness in such a case would have to rely on environmental monitoring, changes in biologic markers of exposure, or some other indirect measure of impact on disease incidence. Developing good baseline and surveillance information about disease incidence, linking health and environmental data, and determining regional differences in disease prevalence, trends, and risk factors would improve the ability to implement effective public health interventions and enhance our confidence that they are effective.

B15 COLLABORATION

The Commission's framework is intended to be implemented collaboratively, with some level of participation of stakeholders or other affected parties at each stage. Figure B3 shows a particularly prominent role for stakeholder involvement in the first stage, formulating the problem in context. Such partnerships can facilitate the exchange of information and ideas that all parties need if they are to make informed decisions about reducing risks. Regulatory actions or decisions are more likely to be successful if affected parties are involved in scoping and decision making than if they are not (Richards 1993). As NAACP representative Walter Fields testified at the Commission's meeting in New Jersey, if people are not included in risk management decisions, such as facility siting, from the very start, they feel excluded from important decisions that affect their communities, and emotional, not rational, reactions govern their response. The importance of stakeholder participation was emphasized by the President's Council on Sustainable Development (PCSD 1996) and the National Research Council Committee on Risk Characterization (NRC 1996).

Including stakeholders in the processes of defining a problem and assessing its risks can provide a forum in which to clarify the technical data and science policy assumptions used in risk assessment. Our recommendation for serious involvement of stakeholders in active protection of ecologic and human health, especially at the community level, is well-supported by recent public opinion poll results. For example, results from a survey in mid-March 1996 suggest that 80% of Americans think that government at all levels should encourage citizen involvement in health and environmental protection (Council for Excellence in Government 1996). They do not want government to do less about risks to health and the environment than it does currently, but they want government action to be more efficient and effective. They also think that responsibility for controlling risks should be shared by government, business, communities, and individuals.

Collaboration also plays a central role in effective implementation, especially if environmental protection is no longer considered solely a government–industry responsibility and the public is expected to participate directly in implementation of risk management steps (McCallum and Santos 1995). Public actions include reduction and recycling of wastes at home, on farms, in offices, and in recreation, as well as bearing some of the cost. Public comment and public meetings are not adequate substitutes for collaborative approaches to problem solving, although they can be useful in gaining broader participation. Effective stakeholder involvement in regulatory decision making requires a shift in attitudes so that affected members of the public are seen as partners in the problem-solving process, rather than as obstacles to it (Van Horn 1988; Chess *et al.* 1995). It might also require a modification in the timelines that regulatory agencies must satisfy to meet statutory or court-imposed requirements.

A potential disadvantage of our framework might be the investments of time and money required to implement a collaborative, systematic process. Even if the process might lead to long-term savings, the up-front costs could be considerable. Moreover, there is no guarantee that consensus on a risk management decision will be reached. However, the time and expense required are unlikely to exceed the experiences of the Occupational Safety and Health Administration and the EPA, which have sometimes required years in preparing agency risk assessments, in notice-and-comment rule-making, and in litigation over the resulting decisions.

Implementing a collaborative decision-making process should include incentives for participation. An industry or municipality might be more inclined to participate willingly and cooperatively if it was to receive some relief in exchange, such as reduced reporting requirements, suspension of punitive damages, or elimination of parties' abilities to sue after a decision is reached.

B16 ITERATION

The framework is intended to be implemented iteratively; that is, the process can be refined and its conclusions can be changed on the basis of research, new data, and new views. Iteration could apply to a rule that has already been promulgated or to a new rule or a new approach to a problem that is being developed. Public comment, negotiation, or analysis can redefine the problem or identify other issues of concern in a broader context. Research and information-gathering performed to clarify a problem or a risk might lead to a focus on a somewhat different problem. Analyzing risks and options can lead to a better understanding of how a problem should have been defined and scoped at the outset. However, iteration must not be allowed to become a device for indefinite delay. Using an iteration to scope a problem might actually speed up the risk management process, as goals and issues are clarified, and possibly lead to a quicker and more cost-effective resolution than expected initially if it becomes apparent that proceeding with the entire framework is no longer necessary.

B17 USING THE FRAMEWORK

The proposed framework is intended to be a guide for an approach or thought process for risk management decision making. It is unlikely that all aspects of the framework would be required for every problem and some might be inconsistent with certain statutory requirements. Different levels of decision making will require different levels of analysis. Risk managers should apply this process flexibly to accommodate the needs of individual circumstances.

A number of criteria might be used to determine when applying the framework would be most useful. A problem worth addressing according to the framework should involve achieving an agreed-upon goal, the optimal path to which is controversial. The problem should require resolution of inter-dependent or related issues. Enough facts should be available to permit meaningful discussion and resolution. Participants in the process should be representative of those affected by the outcome of problem resolution. In some cases, it might be particularly important for elected officials or their representa-tives to be included in the process so that their support of the result is likely. For example, including members of Congress in the stakeholder deliberations about Superfund reauthorization in 1994 might have enhanced the prospects for the success of that effort.

The framework can be applied at several levels.

B17.1 NATIONAL LEVEL

Congress could apply the integrated, multimedia approach of the framework to its future risk management legislation and to its oversight of existing agency programs. For example, Congress could modify the Clean Water Act to establish a comprehensive, integrated, multimedia watershed management approach and to provide for the development of state watershed programs. The current EPA watershed protection approach could be expanded with additional authorization and funding from Congress to accomplish multimedia environmental risk management under the Clean Water Act and possibly under the Clean Air Act. In fact, using the current watershed protection approach, EPA-sponsored estuary programs in Tampa Bay and Galveston Bay are good examples of how state and local governments and citizens can participate in a process that identifies high-priority, multimedia environmental problems and take action to ameliorate the problems.

The EPA can use the framework to support its development of an integrated air toxics strategy for urban areas, to link decisions on residual risks from major sources with risk driven decisions on smaller stationary sources and mobile sources. The EPA can also use the framework to support its Common Sense Initiative to integrate all the permitting that is required of individual manufacturing facilities, overcoming conflicting and redundant requirements.

B17.2 STATE AND REGIONAL LEVELS

Under existing federal and state laws, states and regional airshed or watershed authorities can use the framework approach to address various environmental problems in an integrated manner, where applicable and feasible. Both the Michigan Department of Environmental Quality and EPA Region 5 have organized themselves into multimedia teams to facilitate integrated approaches to risk management. Several states have initiated programs to resolve and integrate potentially conflicting permitting requirements.

One example of how the problem/context and risks stages of the framework could be applied at the state level is the case of oxygenated and reformulated fuels. The 1990 Clean Air Act Amendments required that new vehicle fuels be introduced by 1992 in communities with carbon monoxide concentrations exceeding national ambient air-quality standards. The new fuels contain the additive methyl tert-butyl ether (MTBE). In some states, particularly Alaska, Montana, Wisconsin, and New Jersey, there have been numerous complaints of unexplained symptoms of health effects. More that 70 million Americans are potentially exposed to evaporative emissions from oxygenated and reformu-lated fuels. MTBE has been singled out as potentially responsible for the symptoms. Assessing potential risks associated with these fuels in a public health context should include evaluating not only MTBE toxicity but also risks associated with increased concentrations of nitrogen oxides, formaldehyde, and acetaldehyde. The health benefits of the MTBE-related reductions in carbon monoxide and benzene concentrations would also have to be considered, as would the role of ambient temperature and engine performance in exposures (Health Effects Institute 1996). The risk characterization should also consider the different strengths of evidence for health risks associated with carbon monoxide, benzene, and MTBE. Using a common metric to compare the health effects of each chemical, such as a margin-of-exposure approach, would be useful.

The following is an example of the framework in action at the regional level.

Beginning in 1978, the Association of Bay Area Governments (ABAG) developed and adopted an integrated, comprehensive environmental manage-ment plan for the San Francisco Bay region (ABAG 1978). The plan recom-mended actions to improve and manage the region's air quality, water quality, water supply, and solid waste problems in an integrated, comprehensive manner. ABAG was designated as the lead agency by the state of California under various federal and state laws. The plan was developed to meet the requirements of the

Air Quality Maintenance Plan under the Federal Clean Air Act Amendments of 1977, of the areawide plan under the Federal Water Pollution Control Act Amendments of 1972, of the Federal Resource Conservation and Recovery Act of 1976, of the California Solid Waste Management and Resource Recovery Act of 1972, and of California Senate Bill 424 of 1977.

B17.2.1 Collaboration

The plan was developed through an extensive collaborative process that involved a broad range of stakeholders, including representatives of federal, state, and local regulatory agencies, business, labor, environmental groups, ethnic minorities, and city and county governments. A 46-member environmental management task force charged with plan preparation was formed by ABAG with the stakeholders well represented. The task force was chaired by Dianne Feinstein, then mayor of San Francisco. Community outreach was extensive, and several hundred roundtables, meetings, and formal public hearings were held.

B17.2.2 Problem/context

Stakeholders were involved at the beginning of management plan development to identify environmental problems in the region. For example, photochemical oxidants in the air, toxic materials in the San Francisco Bay, and inadequate solid waste disposal practices were identified as some of the important problems. Potential relationships among the problems were also identified so that they could be considered in a multimedia, integrated manner. For example, solid waste disposal sites could cause surface-water and groundwater contamination and could produce dust, gases, and odors that affected air quality. Therefore, properly managed landfills were considered to be control measures for air and water pollution, as well as for solid waste disposal problems.

B17.2.3 Risks

Risk information was compiled and communicated to the stakeholders. For example, health effects of photochemical oxidants and harmful effects of toxic materials on aquatic life were described to the stakeholders at various meetings. However, risk assessment was not performed extensively, because control measures were chosen on the basis of federal and state requirements.

B17.2.4 Options

Options for control measures were developed with extensive input from stakeholders. The most controversial measure was land-use controls for maintenance of air quality after 1985. The proposed control measures were evaluated for their environmental benefits, consequences, and costs, and for their probable social and economic effects on the region.

B17.2.5 Decisions

Control measures were selected by the environmental management task force, which was composed of locally elected officials and other stakeholders. Many of the control measures were voted on and adopted by the task force. However, after many months of discussion and the expression of substantial concern by labor, business, and many of the cities and counties in the region, the land-use control measure for maintenance of air quality was eliminated from the plan. In the decision-making process, several important issues were raised, including federal–state–local relationships, the social and economic impact of land-use controls, the extent of air quality improvement likely to be obtained, and the suitability of including these measures in an air quality plan.

B17.2.6 Actions

Because so much care was taken to analyze problems and solutions and to make decisions with broad stakeholder participation, many of the actions recommended by the plan were taken by public agencies, businesses, industries, and private citizens. The plan continues to serve as a blueprint for environmental management activities in the bay region. For example, a state implementation plan for air quality was developed in response to the plan and, as a result, the region was designated as an attainment area for ozone under the federal Clean Air Act in 1995. Almost all the industrial and municipal wastewater treatment facilities have been upgraded. Erosion-control measures to reduce nonpoint-source pollution have been in place for many years. A council of water supply agencies has been formed and has engaged in cooperative efforts, such as the development of a drought-response strategy for the region. Hazardous material spill response teams have become available at the city and county levels. ABAG also provided technical assistance to local agencies to initiate recycling programs.

B17.2.7 Iteration

The plan recommended procedures for continual adjustment as new information or new technologies became available. Iteration has, in fact, been carried out over the years.

B17.2.8 Evaluation

As a result of this collaborative, integrated environmental planning effort, the San Francisco Bay region has enjoyed substantial environmental improvement over the last 15 years. Although ABAG's designation as lead agency has expired, a planning process was continued by state and local agencies. Although no formal evaluation of the plan has been performed, much of the environmental improvement in the region can be attributed to implementation of the plan. The San Francisco Bay region is now considered by many to be one of the environmentally best managed metropolitan areas in the country, as a result of the comprehensive, integrated, collaborative environmental management effort started almost 20 years ago.

B17.3 LOCAL AND COMMUNITY LEVELS

A city or a community can use the framework to address risks to its citizens. For example, radioactive waste cleanup is a community level, multimedia problem and should include effective local stakeholder involvement. When the DOE began to address the cleanup problem at Hanford, the surrounding community was not adequately informed or involved, and that led to outrage and distrust. Involvement of the community as partners in risk management at the site since then has led to improved cooperation and more readily accepted decisions.

Other uses of the framework at the local level include the development and operation of industrial facilities or waste disposal facilities.

B18 SUMMARY

Our risk management framework constitutes a major shift in the emphasis that risk assessment plays in risk management decision making and has three critical advantages. First, it is consistent with using an integrated, holistic, top-down approach to a public health or environmental problem instead of a chemical-by-chemical, medium-by-medium, bottom-up approach to characterizing individual risks; decisions about how to use risk assessment, and how extensively, are made from the perspective of risk management. Second, it emphasizes collaboration, communication, and negotiation in an open and inclusive process among stakeholders so that public values can influence the shaping of risk management strategies; the results are intended to be decisions that are more pragmatic and more easily implemented than those made in the absence of consensus and solutions that incorporate the diversity of interests, knowledge, and technical expertise represented among stakeholders. Third, like the scientific process, the proposed risk management process is iterative; at any stage of the process, the discovery of new information can change conclusions and decisions and lead to reformulation and re-evaluation of the problem at hand.

The Commission emphasizes that the proposed framework will need to be refined with experience. As illustrated by the examples described above, some elements of the framework, such as stakeholder involvement and multimedia analysis, have been tried. However, no risk management effort to date has used all aspects of the framework. Many of the questions and concerns associated with this framework will be clarified as it is more widely used. We recommend that evaluation be an integral component of the process.

REFERENCES

Association of Bay Area Governments (1978). *Environmental Management Plan for the San Francisco Bay Area.* Berkeley, CA.

Carnegie Commission on Science, Technology, and Government (1993). *Risk and the Environment: Improving Regulatory Decision-making.* New York, NY.

Chess C, Salomone KL, Hance BJ, Saville A (1995). Results of a national symposium on risk communication: next steps for government agencies. *Risk Analysis* **15**: 115–125.

Council for Excellence in Government (1996). National Public Opinion Survey. Prepared by Peter Hart and Robert Teeter, Inc., Washington, DC.

Health Effects Institute (1996). *The Potential Health Effects of Oxygenates Added to Gasoline: A Review of the Current Literature. A Special Report.* Cambridge, MA.

McCallum DB, Santos S (1995). Participation and Persuasion. A Communication Perspective on Risk Management. In: *Risk Assessment and Management Handbook* (R. Kollara *et al.*, eds). McGraw Hill Inc., New York.

McGinnis JM, Foege WH (1993). Actual causes of death in the United States. *Journal of the American Medical Association* **270**: 2207–2212.

National Academy of Public Administration (1995). *Setting Priorities, Getting Results: A New Direction for EPA.* Washington, DC.

National Commission on the Environment (1992). *Choosing a Sustainable Future.* World Wildlife Federation, Washington, DC.

National Research Council (1983). *Risk Assessment in the Federal Government: Managing the Process.* National Academy Press, Washington, DC.

National Research Council (1994). *Science and Judgment in Risk Assessment.* National Academy Press, Washington, DC.

National Research Council (1996). *Understanding Risk. Informing Decisions in a Democratic Society.* National Academy Press, Washington, DC.

President's Council on Sustainable Development (1996). *Sustainable America. A New Consensus for Prosperity, Opportunity, and a Healthy Environment for the Future.* Washington, DC.

Richards M (1993). Siting Industrial Facilities. Lessons from the Social Science Literature. Presented at the Fifth Annual International Conference of the Society for the Advancement of Socio-Economics. Environmental Decision-Making. New York City, March 26–28.

Ruckelshaus WD (1995). Stopping the Pendulum. *The Environmental Forum* Nov./Dec., 25–29.

Van Horn CE (1988). *Breaking the Environmental Gridlock.* The Eagleton Institute of Politics, Rutgers University, New Brunswick, NJ.

Appendix C

Proposed Guidelines for Carcinogen Risk Assessment

ENVIRONMENTAL PROTECTION AGENCY

Federal Register (1996), 61, 17959–17963.

C1 SUPPLEMENTARY INFORMATION

In 1983, the National Academy of Sciences (NAS)/National Research Council (NRC) published its report entitled *Risk Assessment in the Federal Government: Managing the Process* (NRC 1983). In that report, the NRC recommended that federal regulatory agencies establish 'inference guidelines' to ensure consistency and technical quality in risk assessments and to ensure that the risk assessment process was maintained as a scientific effort separate from risk management. The 1986 cancer guidelines were issued on September 24, 1986 (51 FR 33992). The *Proposed Guidelines* published today continue the guidelines development process. These guidelines set forth principles and procedures to guide the Environmental Protection Agency (EPA) scientists in the conduct of agency cancer risk assessments and to inform agency decision makers and the public about these procedures.

Both the 1986 guidelines and the current proposal contain inference guidance in the form of default inferences to bridge gaps in knowledge and data. Research conducted in the past decade has elucidated much about the nature of carcinogenic processes and continues to provide new information. The intent of this proposal is to take account of knowledge available now and to provide flexibility for the future in assessing data and employing default inferences, recognizing that the guidelines cannot always anticipate future research findings. Because methods and knowledge are expected to change more rapidly than guidelines can practicably be revised, the agency will update specific assessment procedures with peer-reviewed supplementary, technical documents as needed. Further revision of the guidelines themselves will take place when extensive changes are necessary.

Since 1986, the EPA has sponsored several workshops about revising the cancer guidelines (US EPA 1989a,b, 1994a). The Society for Risk Analysis conducted a workshop on the subject in connection with its 1992 annual meeting (Anderson *et al.* 1993). Participants in the most recent workshop in

This appendix is a US Government work and, as such, is in the public domain in the United States of America.

1994 reviewed an earlier version of the guidelines proposed here and made numerous recommendations about individual issues as well as broad recommendations about explanations and perspectives that should be added. Most recently, the Committee on the Environment and Natural Resources of the Office of Science and Technology Policy reviewed the guidelines at a meeting held on August 15, 1995. The EPA appreciates the efforts of all participants in the process and has tried to address their recommendations in this proposal.

C2 MAJOR CHANGES FROM THE 1986 GUIDELINES

C2.1 CHARACTERIZATIONS

Increased emphasis on providing characterization discussions for the hazard, dose–response, and exposure sections is part of the proposal. These discussions will summarize the assessments to explain the extent and weight of evidence, major points of interpretation and rationale, and strengths and weaknesses of the evidence and the analysis, and to discuss alternative conclusions and uncertainties that deserve serious consideration (US EPA 1995). They serve as starting materials for the risk characterization process which completes the risk assessment.

C2.2 WEIGHING EVIDENCE OF HAZARD

A major change is in the way hazard evidence is weighed in reaching conclusions about the human carcinogenic potential of agents. In the 1986 cancer guidelines, tumor findings in animals or humans were the dominant components of decisions. Other information about an agent's properties, its structure–activity relationships to other carcinogenic agents, and its activities in studies of carcinogenic processes was often limited and played only a modulating role as compared with tumor findings. In this proposal, decisions come from considering all of the evidence. This change recognizes the growing sophistication of research methods, particularly in their ability to reveal the modes of action of carcinogenic agents at cellular and subcellular levels as well as toxicokinetic and metabolic processes. The effect of the change on the assessment of individual agents will depend greatly on the availability of new kinds of data on them in keeping with the state of the art. If these new kinds of data are not forthcoming from public and private research on agents, assessments under these guidelines will not differ significantly from assessments under former guidelines.

Weighing of the evidence includes addressing the likelihood of human carcinogenic effects of the agent and the conditions under which such effects may be expressed, as these are revealed in the toxicological and other biologically important features of the agent. (Consideration of actual human exposure and

risk implications are done separately; they are not parts of the hazard characterization). In this respect, the guidelines incorporate recommendations of the
NRC (1994). In that report, the NRC recommends expansion of the former
concept of hazard identification, which rests on simply a finding of carcinogenic potential, to a concept of characterization that includes dimensions of the
expression of this potential. For example, an agent might be observed to be
carcinogenic via inhalation exposure and not via oral exposure, or its carcinogenic activity might be secondary to another toxic effect. In addition, the
consideration of evidence includes the mode(s) of action of the agent apparent
from the available data as a basis for approaching dose–response assessment.

C3 CLASSIFICATION DESCRIPTORS

To express the weight of evidence for carcinogenic hazard potential, the 1986
cancer guidelines provided summary rankings for human and animal cancer
studies. These summary rankings were integrated to place the overall evidence
in classification groups A through E, group A being associated with the greatest
probability of human carcinogenicity and group E with evidence of noncarcino-
genicity in humans. Data other than tumor findings played a modifying role
after initial placement of an agent into a group.

These *Proposed Guidelines* take a different approach, consistent with the
change in the basic approach to weighing evidence. No interim classification of
tumor findings followed by modifications with other data takes place. Instead,
the conclusion reflects the weighing of evidence in one step. Moreover,
standard descriptors of conclusions are employed rather than letter designa-
tions, and these are incorporated into a brief narrative description of their
informational basis. The narrative with descriptors replaces the previous letter
designation. The descriptors are in three categories: 'known/likely', 'cannot be
determined', or 'not likely'. For instance, using a descriptor in context, a
narrative could say that an agent is likely to be carcinogenic by inhalation
exposure and not likely to be carcinogenic by oral exposure. The narrative
explains the kinds of evidence available and how they fit together in drawing
conclusions, and points out significant issues/strengths/limitations of the data
and conclusions. Subdescriptors are used to refine the conclusion further. The
narrative also summarizes the mode of action information underlying a recommended approach to dose–response assessment.

In considering revision of the former classification method, the agency has
examined other possibilities that would retain the use of letter and number
designation of weights of evidence. The use of standard descriptors within a
narrative presentation is proposed for three primary reasons. First, the
proposed method permits inclusion of explanations of data and of their
strengths and limitations. This is more consistent with current policy emphasis
on risk characterization. Second, it would take a large set of individual number

or letter codes to cover differences in the nature of contributing information (animal, human, other), route of exposure, mode of action, and relative overall weight. When such a set becomes large (10–30 codes), it is too large to be a good communication device, because people cannot remember the definitions of the codes so they have to be explained in narrative. Third, it is impossible to predefine the course of cancer research and the kinds of data that may become available. A flexible system is needed to accommodate change in the underlying data and inferences, and a system of codes might become out of date, as has the one in the 1986 cancer guidelines.

C4 DOSE–RESPONSE ASSESSMENT

The approach to dose–response assessment calls for analysis that follows the conclusions reached in the hazard assessment as to potential mode(s) of action. The assessment begins by analyzing the empirical data in the range of observation. When animal studies are the basis of the analysis, the estimation of a human equivalent dose utilizes toxicokinetic data, if appropriate and adequate data are available. Otherwise, default procedures are applied. For an oral dose, the default is to scale daily applied doses experienced for a lifetime in proportion to body weight raised to the 0.75 power. For an inhalation dose, the default methodology estimates respiratory deposition of particles and gases and estimates internal doses of gases with different absorption characteristics. These two defaults are a change from the 1986 cancer guidelines which provided a single scaling factor of body weight raised to the 0.66 power. Another change from the 1986 guidelines is that response data on effects of the agent on carcinogenic processes are analyzed (nontumor data) in addition to data on tumor incidence. If appropriate, the analyses of data on tumor incidence and on precursor effects may be combined, using precursor data to extend the dose–response curve below the tumor data. Even if combining data is not appropriate, study of the dose–response for effects believed to be part of the carcinogenic influence of the agent may assist in thinking about the relationship of exposure and response in the range of observation and at exposure levels below the range of observation.

Whenever data are sufficient, a biologically based or case-specific dose–response model is developed to relate dose and response data in the range of empirical observation. Otherwise, as a standard, default procedure, a model is used to curve-fit the data. The lower 95% confidence limit on a dose associated with an estimated 10% increased tumor or relevant nontumor response LED_{10} is identified. This generally serves as the point of departure for extrapolating the relationship to environmental exposure levels of interest when the latter are outside the range of observed data. The environmental exposures of interest may be measured ones or levels of risk management interest in considering potential exposure control options. Other points of departure may be more appropriate for certain data sets; as described in the guidance, these may be used instead of the LED_{10}. Additionally, the LED_{10} is available for comparison with parallel analyses of other carcinogenic agents or of noncancer effects of agents and for gauging and explaining the magnitude of subsequent extrapolation to low-dose levels. The LED_{10}, rather than the ED_{10} (the estimate of a 10% increased response), is the proposed standard point of departure for two reasons. One is to permit easier comparison with the benchmark dose procedure for noncancer health assessment—also based on the lower limit on dose. Another is that the lower limit, as opposed to the central estimate, accounts for uncertainty in the experimental data. The issue of using a lower limit or central estimate was discussed at a workshop held on the benchmark procedure for noncancer assessment (Barnes *et al.* 1995) and at a workshop on a previous version of this proposal (US EPA 1994b). The latter workshop recommended a central estimate; the benchmark workshop recommended a lower limit.

The second step of dose–response assessment is extrapolation to lower dose levels, if needed. This is based on a biologically based or case-specific model if supportable by substantial data. Otherwise, default approaches are applied that accord with the view of mode(s) of action of the agent. These include approaches that assume linearity or nonlinearity of the dose–response relationship or both. The default approach for linearity is to extend a straight line to zero dose, zero response. The default approach for nonlinearity is to use a margin of exposure analysis rather than estimating the probability of effects at low doses. A margin of exposure analysis explains the biological considerations for comparing the observed data with the environmental exposure levels of interest and helps in deciding on an acceptable level of exposure in accordance with applicable management factors. The use of straight line extrapolation for a linear default is a change from the 1986 guidelines which used the 'linearized multistage' (LMS) procedure. This change is made because the former modeling procedure gave an appearance of specific knowledge and sophistication unwarranted for a default. The proposed approach is also more like that employed by the Food and Drug Administration (FDA 1987). The numerical results of the straight line and LMS procedures are not significantly different (Krewski *et al.* 1984). The use of a margin of exposure approach is included as a new default procedure to accommodate cases in which there is sufficient evidence of a nonlinear dose–response, but not enough evidence to construct a mathematical model for the relationship. (The agency will continue to seek a modeling method to apply in these cases. If a modeling approach is developed, it will be subject to peer review and public notice in the context of a supplementary document for these guidelines.)

The public is invited to provide comments to be considered in EPA decisions about the content of the final guidelines. After the public comment period, the EPA Science Advisory Board will be asked to review and provide advice on the guidelines and issues raised in comments. EPA asks those who respond to this notice to include their views on the following:

1. The proposed guidance for characterization of hazard, including the weight of evidence descriptors and weight of evidence narrative which are major features of the proposal. There are three categories of descriptors: 'known/likely', 'cannot be determined', and 'not likely', which are further refined by subdescriptors. It is felt that these three descriptors will satisfactorily delineate the types of evidence bearing on carcinogenicity as they are used with subdescriptors in the context of a narrative of data and rationale. However, an issue that has been discussed by external peer reviewers and by EPA staff is whether the descriptor–subdescriptor called 'cannot be determined—suggestive evidence' should become a separate, fourth category called 'suggestive'. The EPA may choose this course in the final guidelines and requests comment. The EPA asks commenters on this question to address the rationale (science as well as policy) for leaving the categories of descriptors as proposed or making the fourth category. How might the coverage of a 'suggestive' category be defined in order to be most useful?

2. The use of mode of action information in hazard characterization and to guide dose–response assessment is a central part of the proposed approach to bringing new research on carcinogenic processes to bear in assessments of environmental agents. The appropriate use of this information now and in the future is important. The EPA requests comment on the treatment of such information in the proposal, including reliance on peer review as a part of the judgmental process on its application.

3. Uses of nontumor data in the dose–response assessment and the methodological and science policy issues posed are new to these guidelines. The EPA requests comment on both issues.

4. Dose–response assessment is proposed to be considered in two parts: range of observed data and range of extrapolation. The lower 95% confidence limit on a dose associated with a 10% response (tumor or nontumor response) is proposed as a default point of departure, marking the beginning of extrapolation. This is a parallel to the benchmark procedure for evaluating the dose–response of noncancer health endpoints (Barnes *et al.* 1995). An alternative is to use the central estimate of a 10% response. Another alternative is to use a 1%, instead of a 10%, response when the observed data are tumor incidence data. Does the generally larger sample size of tumor effect studies support using a 1% response as compared with using 10% for smaller studies? Are there other approaches for the point of departure that might be considered?

REFERENCES

Anderson E, Deisler PF, McCallum D, St. Helaire C, Spitzer HL, Strauss H, Wilson JD, Zimmerman R (1993). *Key Issues in Carcinogen Risk Assessment Guidelines.* Society for Risk Analysis, McLean, Virginia.

Barnes DG, Daston GP, Evans JS, Jarabek AM, Kavlok RJ, Kimmel CA, Park C, Spitzer HL (1995). Benchmark dose workshop: Criteria for use of a benchmark dose to estimate a reference dose. *Regulatory Toxicology and Pharmacology* **21**: 296–306.

Krewski D, Brown C, Murdoch D (1984). Determining 'safe' levels of exposure: Safety factors of mathematical models. *Fundamentals in Applied Toxicology* **4**: S383–394.

National Research Council (1983). *Risk Assessment in the Federal Government: Managing the Process.* National Academy Press, Washington, DC.

National Research Council (1994). *Science and Judgment in Risk Assessment.* National Academy Press, Washington, DC.

US Environmental Protection Agency (1989a). *Workshop on EPA Guidelines for Carcinogen Risk Assessment.* EPA/625/3-89/015. Risk Assessment Forum, Washington, DC.

US Environmental Protection Agency (1989b). *Workshop on EPA Guidelines for Carcinogen Risk Assessment: Use of Human Evidence.* EPA/625/3-90/017. Risk Assessment Forum, Washington, DC.

US Environmental Protection Agency (1994a). Estimating Exposure to Dioxin-like Compounds. External Review Draft, 3 vol. EPA/600/6-88/005 Ca, Cb, Cc. Office of Health and Environmental Assessment, Office of Research and Development, Washington, DC.

US Environmental Protection Agency (1994b). *Report on the Workshop on Cancer Risk Assessment Guidelines Issues.* EPA/630/R-94/005a. Office of Research and Development, Washington, DC.

US Environmental Protection Agency (1995). Policy for Risk Characterization. Memorandum of Carol M. Browner, Administrator, March 21, 1995. Washington, DC.

US Food and Drug Administration (1987). Sponsored Compounds in Food-producing Animals; Criteria and Procedures for Evaluating the Safety of Carcinogenic Residues. Final Rule. 21 CFR Parts 70, 500, 514 and 571.

Appendix D

Soil Screening Guidance: Fact Sheet

OFFICE OF SOLID WASTE AND EMERGENCY RESPONSE,
OFFICE OF EMERGENCY AND REMEDIAL RESPONSE,
UNITED STATES ENVIRONMENTAL PROTECTION AGENCY

Publication 9355.4-14FSA PB96-963501. July 1996 EPA/540/F-95/041

This fact sheet summarizes key aspects of the US Environmental Protection Agency's (EPA) Soil Screening Guidance. The Soil Screening Guidance is a tool developed by the EPA to help standardize and accelerate the evaluation and cleanup of contaminated soils at sites on the National Priorities List (NPL) where future residential land use is anticipated. The User's Guide (US EPA 1996a) provides a simple step-by-step methodology for environmental science/engineering professionals to calculate risk-based, site-specific soil screening levels (SSLs) for contaminants in soil that may be used to identify areas needing further investigation at NPL sites. The Technical Background Document (TBD) (US EPA 1996b) presents the analysis and modeling upon which this approach is based, as well as generic SSLs calculated using conservative default values, and guidance for conducting more detailed analysis of complex site conditions, where needed.

D1 SOIL SCREENING LEVELS ARE NOT NATIONAL CLEANUP STANDARDS

SSLs alone do not trigger the need for response actions or define 'unacceptable' levels of contaminants in soil. In this guidance, 'screening' refers to the process of identifying and defining areas, contaminants, and conditions, at a particular site that do not require further federal attention. Generally, at sites where contaminant concentrations fall below SSLs, no further action or study is warranted under the Comprehensive Environmental Response, Compensation and Liability Act (CERCLA), commonly known as the 'Superfund'. (Some states have developed screening numbers that are more stringent than the generic SSLs presented here; therefore, further study may be warranted under state programs.) Where contaminant concentrations equal or exceed SSLs, further study or investigation, but not necessarily cleanup, is warranted.

This appendix is a US Government work and, as such, is in the public domain in the United States of America.

The decision to use the Soil Screening Guidance at a site will be driven by the potential benefits of eliminating areas, exposure pathways, or contaminants from further investigation. By identifying areas where concentrations of contaminated soil are below levels of concern under CERCLA, the guidance provides a means to focus resources on exposure areas, contaminants, and exposure pathways of concern.

SSLs are risk-based concentrations derived from standardized equations combining exposure information assumptions with EPA toxicity data. Three options for developing screening levels are included in the guidance, depending on how the numbers will be used to screen at a site, and the amount of site-specific information that will be collected or is available. Details of these approaches are presented in the User's Guide (US EPA 1996a) and the TBD (US EPA 1996b). The three options for using SSLs are:

- applying generic SSLs
- developing simple, site-specific SSLs
- developing site-specific SSLs based on more detailed modeling

The progression from generic to simple site-specific and more detailed site-specific SSLs usually involves an increase in investigation costs and, generally a decrease in the stringency of the screening levels because conservative assumptions can be replaced with less conservative site-specific information. Generally, the decision of which method to use involves balancing the increased investigation costs with the potential savings associated with higher (but protective) SSLs. The User's Guide focuses on the application of a simple site-specific approach by providing a step-by-step methodology to calculate site-specific SSLs. The TBD provides more information about the other approaches.

Generic SSLs for the most common contaminants found at NPL sites are included in the TBD. Generic SSLs are calculated from the same equations presented in the User's Guide, but are based on a number of default assumptions chosen to be protective of human health for most site conditions. Generic SSLs can be used in place of site-specific screening levels; however, in general, they are expected to be more stringent than site-specific levels. The site manager should weigh the cost of collecting the data necessary to develop site-specific SSLs with the potential for deriving a higher SSL that provides an appropriate level of protection.

The TBD also includes more detailed modeling approaches for developing screening levels that take into account more complex site conditions than the simple site-specific methodology emphasized in the User's Guide. More detailed approaches may be appropriate when site conditions (e.g., a thick vadose zone) are different from those assumed in the simple site-specific methodology presented here. The technical details supporting the methodology used in the User's Guide are provided in the TBD. SSLs developed in accordance with the

User's Guide are based on future residential land-use assumptions and related exposure scenarios. Using this guidance for sites where residential land-use assumptions do not apply could result in overly conservative screening levels; however, the EPA recognizes that some parties responsible for sites with non-residential land use might still find benefit in using the SSLs as a tool to conduct a conservative initial screening.

SSLs developed in accordance with this guidance could also be used for Resource Conservation and Recovery Act (RCRA) corrective action sites as 'action levels', as the RCRA corrective action program currently views the role of action levels as generally fulfilling the same purpose as SSLs. In addition, states may use this guidance in their voluntary cleanup programs, to the extent they deem appropriate. When applying SSLs to RCRA corrective action sites or for sites under state voluntary cleanup programs, users of this guidance should recognize, as stated above, that SSLs are based on residential land-use assumptions. Where these assumptions do not apply, other approaches for determining the need for further study might be more appropriate.

D1.1 ROLE OF SOIL SCREENING LEVELS

In identifying and managing risks at contaminated sites, the EPA considers a spectrum of contaminant concentrations. The level of concern associated with those concentrations depends on the likelihood of exposure to soil contamination at levels of potential concern to human health or to ecological receptors.

Figure D1 illustrates the spectrum of soil contamination encountered at Superfund sites and the conceptual range of risk management responses. At one end are levels of contamination that clearly warrant a response action; at the other end are levels that are below regulatory concern. Screening levels identify the lower bound of the spectrum—levels below which there is generally no concern under CERCLA, provided conditions associated with the SSLs are met. Appropriate cleanup goals for a particular site may fall anywhere within this range depending on site-specific conditions.

The EPA anticipates the use of SSLs as a tool to facilitate prompt identification

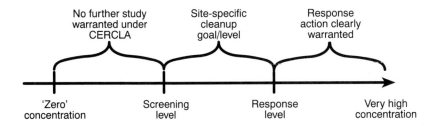

Figure D1 Conceptual risk management spectrum for contaminated soil.

of contaminants and exposure areas of concern during both remedial actions and some removal actions under CERCLA. However, the application of this or any screening methodology is not mandatory at sites being addressed under CERCLA or RCRA. The framework leaves discretion to the site manager and technical experts (e.g., risk assessors, hydrogeologists) to determine whether a screening approach is appropriate for the site and, if screening is to be used, the proper method of implementation. The decision to use a screening approach should be made early in the process of investigation at the site.

The EPA developed the Soil Screening Guidance to be consistent with and to enhance the current Superfund investigation process and anticipates its primary use during the early stages of a remedial investigation (RI) at NPL sites. It does not replace the RI/Feasibility Study (RI/FS), including the risk assessment portion of the RI, but the use of screening levels can focus sampling and risk assessment on aspects of the site that are likely to be a concern under CERCLA. By screening out areas of sites, potential chemicals of concern, or exposure pathways from further investigation, site managers and technical experts can limit the scope of the field investigation or risk assessment.

SSLs can save resources by helping to determine which areas do not require additional federal attention early in the process. Furthermore, data gathered during the soil screening process can be used in later Superfund phases, such as the baseline risk assessment, feasibility study, treatability study, and remedial design. This guidance may also be appropriate for use by the removal program when demarcation of soils above residential risk-based numbers coincides with the purpose and scope of the removal action.

The simple, site-specific SSLs are likely to be most useful where it is difficult to determine whether areas of soil are contaminated to an extent that warrants further investigation or response (e.g., whether areas of soil at an NPL site require further investigation under CERCLA through an RI/FS). As noted above, the screening levels have been developed assuming residential land use. Although some of the models and methods presented in this guidance could be modified to address exposures under other land uses, the EPA has not yet standardized assumptions for exposure scenarios related to those other uses.

This guidance provides the information needed to calculate SSLs for 110 chemicals. Sufficient information may not be available to develop SSLs for additional chemicals. These chemicals should not be screened out, but should be addressed in the baseline risk assessment for the site. The Risk Assessment Guidance for Superfund (RAGS), Volume 1: Human Health Evaluation Manual (HHEM). Part A, Interim Final, (US EPA 1989a) provides guidelines on conducting baseline risk assessments for NPL sites. In addition, the baseline risk assessment should address the chemicals, exposure pathways, and areas at the site that are not screened out.

Although SSLs are 'risk-based', they do not eliminate the need to conduct a site-specific risk assessment for those areas identified as needing further investigation. SSLs are concentrations of contaminants in soil that are designed to be protective of exposures in a residential setting. A site-specific risk assessment is an evaluation of the risk posed by exposure to site contaminants in various media. To calculate SSLs, the exposure equations and pathway models are run in reverse to backcalculate an 'acceptable level' of a contaminant in soil. For the ingestion, dermal, and inhalation pathways, toxicity criteria are used to define an acceptable level of contamination in soil, based on a one-in-a-million (10^{-6}) individual excess cancer risk for carcinogens and a hazard quotient (HQ) of 1 for noncarcinogens. SSLs are backcalculated for migration to groundwater pathways using groundwater concentration limits: nonzero maximum contaminant level goals (MCLGs), maximum contaminant levels (MCLs), or health-based limits (HBLs) (10^{-6} cancer risk or a HQ of 1) where MCLs are not available.

SSLs can be used as Preliminary Remediation Goals (PRGs) provided appropriate conditions are met (i.e., conditions found at a specific site are similar to conditions assumed in developing the SSLs). The concept of calculating risk-based contaminant levels in soils for use as PRGs (or 'draft' cleanup levels) was introduced in RAGS HHEM Part B, Development of Risk Based Preliminary Remediation Goals (EPA 1991b).

PRGs may then be used as the basis for developing final cleanup levels based on the nine-criteria analysis described in the National Contingency Plan [Section 300.430 (3)(2)(I)(A)]. The directive entitled *Role of the Baseline Risk Assessment in Superfund Remedy Selection Decisions* (EPA 1991c) discusses the modification of PRGs to generate cleanup levels. The SSLs should only be used as cleanup levels when a site-specific nine-criteria evaluation of the SSLs as PRGs for soils indicates that a selected remedy achieving the SSLs is protective, complies with 'applicable or relevant and appropriate requirements' (ARARs), and appropriately balances tradeoffs between cleanup options with respect to the other criteria, including cost.

D1.2 SCOPE OF SOIL SCREENING GUIDANCE

In a residential setting, potential pathways of exposure to contaminants in soil are as follows (see Figure D2):
- direct ingestion
- inhalation of volatiles and fugitive dusts
- ingestion of contaminated groundwater caused by migration of chemicals through soil to an underlying potable aquifer
- dermal absorption
- ingestion of homegrown produce that has been contaminated via plant uptake
- migration of volatiles into basements

The Soil Screening Guidance addresses each of these pathways to the greatest extent practical. The first three pathways—direct ingestion, inhalation of volatiles and fugitive dusts, and ingestion of potable groundwater—are the

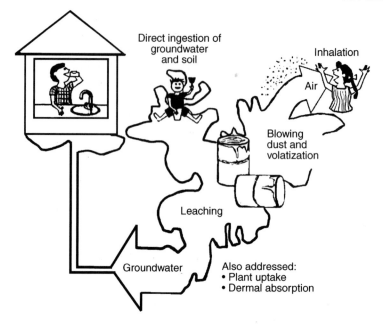

Direct ingestion of
groundwater
and soil

Inhalation

Air

Blowing
dust and
volatization

Leaching

Groundwater

Also addressed:
• Plant uptake
• Dermal absorption

Figure D2 Exposure pathways addressed by soil screening levels (SSLs).

most common routes of human exposure to contaminants in the residential setting. These pathways have generally accepted methods, models, and assumptions that lend themselves to a standardized approach. The additional pathways of exposure to soil contaminants, dermal absorption, plant uptake, and migration of volatiles into basements, may also contribute to the risk to human health from exposure to specific contaminants in a residential setting. The guidance addresses these pathways to a limited extent based on available empirical data. See step 5 in Section D2.5 and the TBD (US EPA 1996b) for further discussion.

The Soil Screening Guidance addresses the human exposure pathways listed previously and will be appropriate for most residential settings. The presence of additional pathways or unusual site conditions does not preclude the use of SSLs in areas of the site that are currently residential or likely to be residential in the future. However, the risks associated with additional pathways or conditions (e.g., fish consumption, raising of livestock, heavy truck traffic on unpaved roads) should be considered in the RI/FS to determine whether SSLs are adequately protective.

An ecological assessment should also be performed as part of the RI/FS to evaluate potential risks to ecological receptors.

The Soil Screening Guidance should not be used for areas with radioactive contaminants.

Key attributes of the user's guide

- Standardized equations are presented to address human exposure pathways in a residential setting consistent with the Superfund's concept of 'reasonable maximum exposure' (RME)

- Source size (area and depth) can be considered on a site-specific basis using mass-limit models.

- Parameters are identified for which site-specific information is needed to develop SSLs.

- Default values are provided to calculate generic SSLs when site-specific information is not available.

- SSLs are based on a 10^{-6} excess risk for carcinogens or a hazard quotient of 1 for noncarcinogens. SSLs for migration to groundwater are based on (in order of preference): nonzero maximum contaminant level goals (MCLGs), maximum contaminant levels (MCLs), or the aforementioned risk-based targets.

Figure D3   Key attributes of the User's Guide. SSL, soil screening levels.

Figure D3 provides key attributes of the *Soil Screening Guidance: User's Guide* (see also US EPA 1996a).

D2 SOIL SCREENING PROCESS

Applying site-specific screening levels involves developing a conceptual site model (CSM), collecting a few easily obtained site-specific soil parameters (such as the dry bulk density and percent moisture), and sampling to measure contaminant concentrations in surface and subsurface soils. Often, much of the information needed to develop the CSM can be derived from previous site investigations, e.g., the Preliminary Assessment/Site Inspection (PA/SI), and, if properly planned, SSL sampling can be accomplished in one mobilization. This fact sheet provides a brief overview of the steps in the process. A full discussion of the steps and their implementation is available in the User's Guide.

The soil screening process (outlined in Figure D4) is a step-by-step approach that involves:

- developing a CSM
- comparing the CSM with the SSL scenario
- defining data collection needs
- sampling and analyzing soils at site
- deriving site-specific SSLs, as appropriate
- comparing site soil contaminant concentrations with SSLs
- determining which areas of the site require further study

The overall outline is fundamentally the same, whether you are using the simple site-specific approach, the generic levels, or a more detailed approach. However, the details of any specific application will be different. In particular, developing the simple site-specific SSLs is obviously more involved than using the generic screening levels available in the TBD.

However, developing site-specific levels may be worthwhile given the less stringent but equally protective levels that will generally result.

An important part of this guidance is a recommended sampling approach that balances the need for more data to reduce uncertainty with the need to limit data collection costs. Where data are limited such that use of the 'maximum test' (Max test) presented in the User's Guide is not appropriate, the guidance also provides direction on the use of other conservative estimates of contaminant concentrations for comparison with the SSLs.

D2.1 STEP 1: DEVELOPING A CONCEPTUAL SITE MODEL

The CSM is a three-dimensional 'picture' of site conditions that illustrates contaminant distributions, release mechanisms, exposure pathways and migration routes, and potential receptors. The CSM documents current site conditions and is supported by maps, cross-sections, and site diagrams that illustrate human and environmental exposure through contaminant release and migration to potential receptors. Developing an accurate CSM is critical to proper implementation of the Soil Screening Guidance.

As a key component of the RI/FS and EPA's Data Quality Objectives (DQO) process, the CSM should be updated and revised as investigations produce new information about a site. A general discussion about the development and use of the CSM during RIs can be found in EPA (1989b, 1993a).

Soil screening process

Step One: **Develop conceptual site model**
- Collect existing site data (historical records, aerial photographs, maps, PA/SI data, available background information, State soil surveys, etc.)
- Organize and analyze existing site data
  - Identify known sources of contamination
  - Identify affected media
  - Identify potential migration routes, exposure pathways, and receptors
- Construct a preliminary diagram of the CSM
- Perform site reconnaissance
  - Confirm and/or modify CSM
  - Identify remaining data gaps

Step two: **Compare soil component of CSM to soil screening scenario**
- Confirm that future residential land use is a reasonable assumption for the site
- Identify pathways present at the site that are addressed by the guidance
- Identify additional pathways present at the site not addressed by the guidance
- Compare pathway-specific generic SSLs with available concentration data
- Estimate whether background levels exceed generic SSLs

Step three: **Define data collection needs for soils to determine which site areas exceed SSLs**
- Develop hypothesis about distribution of soil contamination (i.e., which areas of the site have soil contamination that exceed appropriate SSLs?)
- Develop sampling and analysis plan for determining soil contaminant concentrations
  - Sampling strategy for surface soils (includes defining study boundaries, developing a decision rule, specifying limits on decision errors, and optimizing the design)
  - Sampling strategy for subsurface soils (includes defining study boundaries, developing a decision rule, specifying limits on decision errors, and optimizing the design)
  - Sampling to measure soil characteristics (bulk density, moisture content, organic carbon content, porosity, pH)
- Determine appropriate field methods and establish QA/QC protocols

Step four: **Sample and analyze soils at site**
- Identify contaminants
- Delineate area and depth of sources
- Determine soil characteristics
- Revise CSM, as appropriate

Step five: **Derive site-specific SSLs, if needed**
- Identify SSL equations for relevant pathways
- Identify chemical of concern for dermal exposure and plant uptake
- Obtain site-specific input parameters from CSM summary
- Replace variables in SSL equations with site-specific data gathered in Step 4
- Calculate SSLs
  - Account for exposure to multiple contaminants

Step six: **Compare site soil contaminant concentrations to calculate SSLs**
- For surface soils, screen out exposure areas where all composite samples do not exceed SSLs by a factor of 2
- For subsurface soils, screen out source areas where the highest average soil core concentration does not exceed the SSLs
- Evaluate whether background levels exceed SSLs

Step seven: **Decide how to address areas identified for further study**
- Consider likelihood that additional areas can be screened out with more data
- Integrate soil data with other media in the baseline risk assessment to estimate cumulative risk at the site
- Determine the need for action
- Use SSLs as PRGs

Figure D4 Soil screening process. PA/SI, Preliminary Assessment/Site Inspection; CSM, conceptual site model; SSL, soil screening levels; QA/QC, quality assurance/quality control; PRGs, Preliminary Remediation Goals.

D2.2 STEP 2: COMPARING THE CONCEPTUAL SITE MODEL WITH SOIL SCREENING LEVELS SCENARIO

In this step, the CSM for a particular site is compared with the CSM assumed for the development of the Soil Screening Guidance. This comparison should determine whether the SSL scenario is sufficiently similar to the CSM so that use of the guidance is appropriate. The Soil Screening Guidance was developed assuming residential land use. The primary exposure pathways associated with

---

Residential exposure assumptions

Exposure frequency.................350 days/year
Exposure duration...........................30 years

<u>For noncarcinogens</u>

Body weight...15 kg
Ingestion rate...............................200 mg/day

<u>For carcinogens</u>

Body weight.....age adjusted from 15 to 70 kg
Ingestion rate......................age adjusted from
                              200 to 100 mg/day
Drinking water ingestion rate...............2 l/day
Inhalation rate...............................20 m^3/day

Figure D5   Residential exposure assumptions.

residential land use (given in Section D1.2) are: (1) direct ingestion; (2) inhalation of volatile and fugitive dusts; and (3) ingestion of contaminated groundwater caused by migration of chemicals through soil to an underlying potable aquifer. The residential exposure assumptions associated with these pathways are given in Figure D5.

The CSM may include other sources and exposure pathways that are not covered by this guidance. Compare the CSM with the assumptions and limitations inherent in the SSLs to determine whether additional or more detailed assessments are needed for any exposure pathways or chemicals. The Soil Screening Guidance can be used to screen those sources and exposures pathways that are covered by the guidance. Early identification of areas or conditions where SSLs are not applicable is important so that other characterization and response efforts can be considered when planning the sampling strategy.

Where the following conditions exist, a more detailed site-specific investigation will be needed:

- site adjacent to surface water
- potential terrestrial or aquatic ecological concerns
- other human exposure pathways likely (e.g., local fish consumption, homegrown dairy, livestock or other agricultural use, or

- unusual site conditions (e.g., presence of nonaqueous phase liquids, unusually high fugitive dust levels from site activities)

A consideration of background concentrations should be made to determine whether SSLs are likely to be useful, as the SSLs have much less utility where background concentrations exceed the SSLs. Background concentrations exceeding generic SSLs do not necessarily indicate that a health threat exists, but may suggest that additional analysis is appropriate. For example, it may be important to determine whether the high background concentrations are anthropogenic or naturally occurring. Generally, the EPA does not clean up below natural background; however, where anthropogenic background levels exceed SSLs, EPA may determine that some type of comprehensive response is necessary and feasible.

D2.3 STEP 3: DEFINING DATA COLLECTION NEEDS FOR SOILS

Once the CSM has been developed and the site manager has determined that the Soil Screening Guidance is appropriate to use at a site, a Sampling and Analysis Plan (SAP) should be developed. Figure D4 outlines the general strategy for developing sampling plans likely to be needed to apply the Soil Screening Guidance. A different sampling approach is used for the surface and subsurface because different exposure pathways are being addressed. Sampling should also provide site characteristics data necessary to develop site-specific SSLs. The User's Guide provides information on the development of SAPs for these three types of information.

To develop sampling strategies that will properly assess site contamination, the EPA recommends that site managers consult with the technical experts in their region, including risk assessors, toxicologists, chemists, and hydro-geologists, who can assist the site manager to use the DQO process to satisfy Superfund program objectives. The DQO process is a systematic planning process developed by the EPA to ensure that sufficient data are collected to support EPA decision making. A full discussion of the DQO process is provided in EPA (1993a, 1994). Many of the key elements have been incorporated as part of the guidance.

One of the critical decisions to make before developing the SAP is to define the specific area to which the Soil Screening Guidance will be applied. Existing data (e.g., preliminary assessment, other site investigation data, historical documents discussing site activities) can be used to determine what level and type of investigation may be appropriate. Areas known to be important sources of groundwater contamination should be sampled for subsurface contamination, but it often will not be necessary to develop screening levels based on surface contamination for these areas. Sampling in known source areas will focus on developing remedial alternatives with some sampling to confirm expected problems, as necessary. Other areas may have good historical information to indicate that no waste handling activities occurred there and it is expected that these areas are unlikely to be contaminated. A few samples may be taken to confirm this hypothesis. Much of the sampling effort for soil screening is likely to focus on areas of uncertain contamination levels and history. The User's Guide provides more information about the use of historical information, the statistical basis for the sampling strategy, and the soil characteristics that are needed to develop site-specific screening levels.

D2.4 STEP 4: SAMPLING AND ANALYZING SITE SOILS

Once the sampling strategies have been developed and implemented, the samples should be analyzed according to the analytical laboratory and field methods specified in the SAP. An important outcome of these analyses is the estimation of the concentrations of potential contaminants of concern which will be compared with the SSLs. At this point, the generic SSLs may be useful for comparison purposes. Where estimated concentrations are above the generic SSLs, site-specific SSLs can be calculated to provide another, less stringent but still conservative comparison.

Because these analyses reveal new information about the site, update the CSM accordingly.

D2.5 STEP 5: CALCULATING SITE-SPECIFIC SOIL SCREENING LEVELS

With the soil properties data collected in step 4 of the screening process, site-specific SSLs can now be calculated using the equations presented in the User's Guide. The Soil Screening Guidance provides the equations necessary to develop a simple site-specific SSLs. For a description of how these equations were developed, as well as background on their assumptions and limitations, consult the TBD. When generic SSLs are being used as for comparison with site concentration, this step may be omitted.

All SSL equations were developed to be consistent with reasonable maximum exposure (RME) for the residential setting. The Superfund program estimates the RME for chronic exposures on a site-specific basis by combining an average exposure-point concentration with reasonably conservative values for intake and duration (US EPA 1989a, 1991a). Thus, all site-specific parameters (soil, aquifer, and meteorologic parameters) used to calculate SSLs should reflect average or typical site conditions in order to calculate average exposure concentrations at the site.

Exposure pathways addressed in the process for screening surface soils include direct ingestion, dermal contact, and inhalation of fugitive dusts. While the guidance provides all the relevant toxicity from EPA sources necessary to calculate site-specific SSLs, the Integrated Risk Information System (IRIS) (US EPA 1995a) or Health Effects Assessment Summary Tables (HEAST) (EPA l995b) should be checked for current values. Only the most current values should be used to calculate SSLs.

The Soil Screening Guidance addresses two exposure pathways for subsurface soils: inhalation of volatiles and ingestion of groundwater contaminated by the migration of contaminants through soil to an underlying potable aquifer. Because the equations developed to calculate SSLs for these pathways assume an infinite source, they can violate mass-balance considerations, especially for small sources. To address this concern, the guidance also includes equations for calculating mass-limit SSLs for each of these pathways when the size (i.e., area and depth) of the contaminated soil source is known or can be estimated with confidence.

The Soil Screening Guidance uses a simple linear equilibrium soil/water partition equation or a leach test to estimate contaminant release in soil leachate. It also uses a simple water-balance equation to calculate a dilution factor to account for reduction of soil leachate concentration from mixing in an aquifer.

The methodology for developing SSLs for the migration to groundwater pathway was designed for use during the early stages of a site evaluation when information about subsurface conditions may be limited. Hence, the methodology is based on rather conservative, simplified assumptions about the release and transport of contaminants in the subsurface (Figure D6). These assumptions are inherent in the SSL equations and should be reviewed for consistency with the CSM (see step 2) to determine the applicability of SSLs to the migration to groundwater pathway.

D2.5.1 Address exposure to multiple chemicals

The SSLs generally correspond to a 10^{-6} excess risk level for carcinogens and a hazard quotient of 1 for noncarcinogens. This 'target' hazard quotient is used to calculate a soil concentration below which it is unlikely that sensitive populations will experience adverse health effects. The potential for additive effects has not been 'built in' to the SSLs through apportionment. For carcinogens, the EPA believes that setting a 10^{-6} excess risk level for individual chemicals and pathways generally will lead to cumulative site risks within the 10^{-4} to 10^{-6} risk range for the combinations of chemicals typically found at NPL sites.

For noncarcinogens, there is no widely accepted risk range, and the EPA recognizes that cumulative risks from noncarcinogenic contaminants at a site could exceed the target hazard quotient. However, the EPA also recognizes that noncancer risks should be added only for those chemicals with the same toxic endpoint or mechanism of action.

If more than one chemical detected at a site affects the same target organ (i.e., has the same critical effect as defined by the reference dose methodology), an overall hazard index (HI) for the source (or exposure area) can be

**Simplifying assumptions for the
SSL migration to groundwater pathway**

- Infinite source (i.e., steady-state
  concentrations are maintained over the
  exposure period)

- Uniformly distributed contamination from
  the surface to the top of the aquifer

- No contaminant attenuation (i.e.,
  absorption, biodegradation, chemical
  degradation) in soil

- Instantaneous and linear equilibrium
  soil/water partitioning

- Unconfined, unconsolidated aquifer with
  homogeneous and isotropic hydrologic
  properties

- Receptor well at the downgradient edge
  of the source and screened within the
  plume

- No contaminant attenuation in the aquifer

- No NAPLs present (if NAPLs are
  present, the SSLs do not apply).

Figure D6   Simplifying assumptions for the soil screening level (SSL) migration to
groundwater pathway. NAPL, nonaqueous phase liquid.

calculated. If this HI exceeds 1, further investigation is needed. The guidance
provides a list of target organs for all chemicals with SSLs based on noncarcinogenic effect.

D2.6   STEP 6: COMPARING SITE SOIL CONTAMINANT
CONCENTRATIONS TO CALCULATED SOIL SCREENING LEVELS

Now that the site-specific SSLs have been calculated for the potential contaminants of concern, compare them with the site contaminant concentrations. At
this point, it is reasonable to review the CSM with the actual site data to

confirm its accuracy and the overall applicability of the Soil Screening Guidance.

Thus, for surface soils, the contaminant concentrations in each composite sample from an exposure area are compared with two times the SSL. (When SSL DQOs were developed, two times the SSL was determined to a reasonable upper limit for comparison that would still be protective of human health. Use of this decision rule is appropriate only when the quantity and quality of data are comparable with the levels discussed in the User's Guide. For a complete discussion for the SSL DQOs, see the TBD.) If any composite has concentrations that equal or exceed two times the SSL, the area cannot be screened out, and further study is needed. However, if all composite samples are below two times the SSLs, no further study is needed.

For data sets of lesser quality, the 95% upper confidence level on the arithmetic mean of contaminant soil concentration can be compared directly with the SSLs. The TBD discusses strengths and weaknesses of different calculations of the mean and when they are appropriate for making screening decisions.

As subsurface soils are not characterized to the same extent as surface soils, there is less confidence that the concentrations measured are representative of the entire source. Thus, a more conservative approach to screening is warranted. Because it may not be protective to allow for comparison with values above the SSL, mean contaminant concentrations from each soil boring taken in a source area are compared with the calculated SSLs. Source areas with any mean soil boring contaminant concentration greater than the SSLs generally warrant further consideration. On the other hand, where the mean soil boring contaminant concentrations within a source are all less than the SSLs, that source area is generally screened out.

D2.7 STEP 7: ADDRESSING AREAS IDENTIFIED FOR FURTHER STUDY

Areas that have been identified for further study become a subject of the RI/FS (US EPA 1989b). The results of the baseline risk assessment conducted as part of the RI/FS will establish the basis for taking remedial action. The threshold for taking action differs from the criteria used for screening. Remedial action at NPL sites is generally warranted where cumulative risks for current or future land use exceed 1×10^{-4} for carcinogens or an HI of 1 for noncarcinogens (US EPA 1991c). The data collected for soil screening are useful in the RI and baseline risk assessment. However, additional data will probably need to be collected during future site investigations. These additional data will better define the risks and threats at the site and could conceivably indicate that no action is required.

Once the decision has been made that remedial action may be appropriate, the SSLs can then serve as PRGs. This process is referenced in Section D1.1 of this document.

D3 FOR FURTHER INFORMATION

The technical details (e.g., equations and assumptions) necessary to implement the soil screening guidance are available in the *Soil Screening Guidance: User's Guide* (US EPA 1996a). More detailed discussions of the technical background and assumptions supporting the development of the Soil Screening Guidance are presented in the *Soil Screening Guidance: Technical Background Document* (US EPA 1996b). The final portion of the guidance package is the *Soil Screening Guidance: Response to Comments* (US EPA 1996c) which describes changes made to the guidance following peer review and public comment. For additional copies of this fact sheet, the User's Guide, the Technical Background Document, Response to Comments, or other EPA documents, call the National Technical Information Service (NTIS) at (703) 487-4650 or 1-800-553-NTIS (6847).

REFERENCES

US Environmental Protection Agency (1989a). *Risk Assessment Guidance for Superfund,* Vol. 1. *Human Health Evaluation Manual* (Part A). Interim Final. EPA/540/1-89/002. Office of Emergency and Remedial Response, Washington, DC.

US Environmental Protection Agency (1989b). *Guidance for Conducting Remedial Investigations and Feasibility Studies under CERCLA.* EPA/540/G-89/004. OSWER Directive 9355.3-01. NTIS PB89-184626. Office of Emergency and Remedial Response, Washington, DC.

US Environmental Protection Agency (1991a). *Human Health Evaluation Manual, Supplemental Guidance: Standard Default Exposure Factors.* OSWER Directive 9285.6-03. Office of Emergency and Remedial Response, Washington, DC.

US Environmental Protection Agency (1991b). *Risk Assessment Guidance for Superfund,* Vol. 1. *Human Health Evaluation Manual* (Part B). EPA 9285.7-O1B. Office of Emergency and Remedial Response, Washington, DC.

US Environmental Protection Agency (1991c). *Role of the Baseline Risk Assessment in Superfund Remedy Selection Decisions.* Publication 9355.0-30. NTIS PB91-921359/CCE. Office of Emergency and Remedial Response, Washington, DC.

US Environmental Protection Agency (1993a). *Data Quality Objectives for Superfund.* Interim Final Guidance. Publication 9255.9-01. EPA 540-R-93-071. NTIS PB94-963203. Office of Emergency and Remedial Response, Washington, DC.

US Environmental Protection Agency (1993b). *Quality Assurance for Superfund Environmental Data Collection Activities.* Quick Reference Fact Sheet. NTIS PB93-963273. Office of Emergency and Remedial Response, Washington, DC.

US Environmental Protection Agency (1994). *Guidance for the Data Quality Objectives Process.* EPA QA/G-4. Quality Assurance Management Staff, Office of Research and Development, Washington, DC.

US Environmental Protection Agency (1995a). *Integrated Risk Information System (IRIS).* Cincinnati, OH.

US Environmental Protection Agency (1995b). *Health Effects Assessment Summary Tables, FY-1993.* Annual. Environmental Criteria and Assessment Office, Office of Health and Environmental Assessment, Office of Research and Development, Cincinnati, OH.

US Environmental Protection Agency (1996a). *Soil Screening Guidance: User's Guide.* EPA/540/R-96/018. Office of Emergency and Remedial Response, Washington, DC.

US Environmental Protection Agency (1996b). *Soil Screening Guidance: Technical Background Document.* EPA/540/R-96/128. Office of Emergency and Remedial Response, Washington, DC.

US Environmental Protection Agency (1996c). *Soil Screening Guidance: Response to Comments.* EPA/540/R-96/019. Office of Emergency and Remedial Response, Washington, DC.

Appendix E

Risk in Environmental Decision Making. A State Perspective.
Working Paper

THE NATIONAL GOVERNORS' ASSOCIATION
April 15, 1994

E1 INTRODUCTION

This section examines the scope and uses of risk in state environmental agencies. The first part examines how pervasive the use of risk-related activities is among states, what the legal basis is for states' use of risk, and how individual state agencies are using risk assessments. The second part of the study examines how risk assessments are used in state cleanup activities at contaminated sites.

The study was requested by the Environmental and Public/Occupational Health Standards Steering Group to assess how pervasive the use of risk assessments is by states. The Steering Group was established by the directors of the Department of Energy (DOE) laboratories in January 1990 to organize a broad, long-term educational outreach and research program focused on better scientific and public understanding of the risks associated with hazardous materials in the environment and the work place. The Steering Group's work is part of an effort to improve public understanding of risk and the importance of cost/benefit analysis in mitigation of risk and also to better understand the health and environmental effects of hazardous materials. The Steering Group is also analyzing the potential for the evolution of risk-based consensus standards into federal and state environmental regulations.

The bulk of the information about state activities for this study was collected by surveying the individual state agencies engaged in risk-related activities. A screening survey was initially sent to directors and commissioners of state agencies concerned with the following areas: air (stationary and mobile sources); natural resources (dam safety, earthquakes); water (water quality, drinking water); hazardous and solid waste management (Resource Conservation and Recovery Act—RCRA); site remediation (site cleanups, underground storage tanks); occupational health and safety; state-wide environmental

This appendix is a US Government work and, as such, is in the public domain in the United States of America.

planning; and 'other'. The purpose of this survey was to identify those state agencies engaged in risk-related activities. The responses indicated that nearly all states believed they were using some form of risk assessment, management, or communication. Based on the results of the screening survey, a more detailed survey instrument was developed. After several revisions, it was field tested in several states. The results of this field test were incorporated into a revised survey instrument.

In all, some 305 surveys were distributed to various environmental agencies from at least every state and the District of Columbia. A total of 164 surveys was returned for a response rate of 53%. Of the 164 returned surveys, 117 respondents or agencies indicated they were involved in risk-related activities. Only six states and the District of Columbia did not return a survey and of the 44 surveys that were returned only two indicated they did not use any form of risk analysis in environmental decision making.

The completed surveys were evaluated to determine completeness and accuracy, entered into a database, and analyzed. Along with the survey data other National Governors' Association reports on Superfund, risk communication, and comparative risk programs were consulted.

E2 OVERVIEW OF STATE RISK ACTIVITIES

Developing a common state perspective on risk assessment is nearly impossible. There are possibly as many varying viewpoints on risk assessment as there are states. Even within a state the concept of what risk assessment is may differ greatly. One reason the concept of risk assessment differs is that the legal and regulatory basis for using risk assessment is often unclear and is rarely consistent among states. A state air regulatory program may use risk assessment aggressively while risk assessment may not be used at all by a neighboring water program. One explanation for this is because of various legislative constraints, attitudes of state administrators and program officials, and the complex nature of a specific state's environmental problems.

In addition, the consideration of risk assessment in state regulatory programs is constantly evolving to meet the changing demands of protecting human health. Research and change in methodology continually provide decision makers with new information. Detection of contaminants has become increasingly more sophisticated as has the modeling to measure their potential health and environmental impact. Consequently, a site that was once considered 'safe' may now have to be reevaluated.

Clearly, risk will play an increasingly important part in state regulatory and decision making concerning the environment. Nearly all states say they use risk to make decisions. Of the 117 respondents from environmental agencies in 42 states who indicated involvement in risk-related activities our survey indicated:

- 41 states use risk assessment
- 39 states are involved in risk communication
- 37 states perform risk management

Of the 117 survey respondents, 105 agency representatives indicated they use risk assessment in some way: 92 agencies are involved in risk communication and 84 agencies are involved in risk management. Hazardous materials and solid waste programs reported using risk assessments the most (31 agencies in 23 states), followed by environmental and public health programs (26 agencies in 24 states). Air, water, and occupational safety and health programs had roughly an equal distribution among the surveyed agencies.

Our survey provided a great deal of information regarding state risk-related activities. After analyzing the survey results it became apparent that further study would be needed in order to understand fully how states are using risk assessments and how they define risk. Where states have developed more sophisticated, innovative risk programs, a more in-depth study could be conducted, such as a workshop or case study, to gain a more complete picture of their risk-related activities.

E2.1 LEGAL BASIS FOR RISK ACTIVITIES

When agencies were asked what is the legal basis for their current risk-related activities, 77 agencies from 39 states reported that state statutes were the legal basis for their risk activities. Twenty-one hazardous materials and solid waste programs and 17 environmental and public health programs reported state statutes as the legal basis for their risk activities. Interestingly enough, we discovered that approximately half—55 agencies (32 states)—reported that a state regulation was the legal basis for their agency's risk-related activities, while the other half—53 agencies (31 states)—said their risk-related activities were not based on state regulations.

Several states indicated their agency's risk activities were started because of public concerns and pressure. Some states also reported that they began their risk activities to comply with occupational safety and health requirements and to ensure the safety of their workers.

In looking to the future it appears that most state activity will focus on developing regulations. Thirty-two agencies from 21 states indicated they would be issuing a state regulation relating to risk in the near future. Only three agencies from two states reported that they would be issuing a statute related to risk in the near future.

E2.2 RISK ASSESSMENTS

According to the National Academy of Sciences, risk assessment is the 'characterization of potential adverse effects of exposures to hazards, including estimates of risk and of uncertainties in measurements, and use of analytical techniques and interpretive models; quantitative risk assessment characterizes the risk in numerical representation' (National Research Council 1983, 1994). According to our survey, the majority of state agencies agree with the National Academy of Sciences' definition of risk assessment.

The survey results showed occupational safety and health programs are more committed to doing risk assessments than any of the other agencies surveyed. Agriculture and water programs also reported a high use of risk assessments. Natural resources programs have the least involvement in risk assessments. The average number of staff conducting risk assessments in the agencies surveyed is a little more than six and one-half full time equivalent units (FTEs).

E2.3 TYPES OF RISK ASSESSMENTS

State risk assessments fall into four broad categories and usually involve multiple departments within an agency. Eighty-eight agencies in 39 states indicated they conduct health risk assessments, performed most frequently by environmental and public health programs and hazardous materials and solid waste programs. Forty-three agencies in 27 states conduct ecological risk assessments, performed by hazardous materials and solid waste programs most frequently. Twenty-one agencies in 14 states reported conducting economic risk assessments. Environmental justice assessments are conducted by 10 agencies in 10 states to assess whether low income, minority areas are being designated for hazardous waste sites more often than more affluent areas. Twenty-seven agencies in 20 states conduct 'other' risk assessments. The 'other' category ranged from fire safety assessments to fish advisories, and the numbers reported conducting these types of assessments were roughly equal across the agencies surveyed.

E2.4 WHERE AGENCIES USE RISK ASSESSMENTS

Although states indicated risk assessment is being used throughout a variety of programs, most states indicated it is used predominantly by hazardous materials and solid waste programs to determine cleanup goals and remedies. However, the activities covered are broad; our survey results indicated 65 agencies representing 30 states reported using risk assessment in solid and hazardous waste management. Fifty-one agencies in 34 states reported using risk assessments in setting soil contamination standards.

Fifty agencies representing 43 states use risk assessments in setting toxic air emissions. Thirty-three agencies in 16 states reported using risk assessments in setting drinking water standards. Twenty-three agencies from 18 states use risk assessments in other types of activities. Twenty agencies in 17 states reported using risk assessments in habitat and land use. Only 14 agencies representing 12 states reported using risk assessments in radioactive waste management.

E2.5 SETTING REGULATORY STANDARDS FOR PERMISSIBLE CONTAMINATION LEVELS

Our survey asked agencies if they set regulatory standards for permissible contamination levels in food, water, and ambient air, how they determine what level of risk is acceptable; i.e., how do they define 'safe enough'? Out of the 70 agencies in 36 states that set regulatory standards, 45 agencies in 25 states reported that they rely on a 1 in 1 000 000 cancer risk for defining 'safe enough' contamination levels. Hazardous materials and solid waste programs reported using this determinant the most. Thirty-four agencies from 22 states reported that they rely on case-by-case decisions. Hazardous materials and solid waste programs and environmental and public health programs indicated the highest reliance on case-by-case decisions. Only 11 agencies reported setting standards based on political guidance from legislature.

E2.6 COMPARATIVE RISK PROGRAMS

Although the survey was designed to collect information on specific state programs, many states are ranking environmental problems across media in terms of the relative risk posed to human health and the environment. This analysis is known as comparative risk. The process combines quantitative risk analysis with public perception in identifying and evaluating environmental issues, comparing them, and ranking their importance. The result of the process is usually a list of environmental issues arranged in order of the relative risk they pose. Typically, the comparative risk analysis is the first phase of a project. The second phase develops risk management strategies and sets priorities for action.

It is important that limited resources be spent wisely in times of tight budgets in government and the private sector. Clearly, priorities need to be set. A comparative risk analysis of environmental problems is the first step in environmental strategic planning and will be the first step in establishing those priorities. Focusing on risk ensures that limited resources will be used to achieve the greatest reduction in overall risk from the environmental problems states face.

Five states and one territory have completed comparative risk rankings: Colorado, Guam, Louisiana, Michigan, Vermont, and Washington. Nine states currently have programs underway: Alabama, Arizona, California, Hawaii, Illinois, Maine, Oregon, Texas, and Utah. Ten states are in the planning stages: Alaska, Arkansas, Florida, Kentucky, Maryland, Minnesota, Missouri, Ohio, Tennessee, and Wisconsin. Of those states with a completed ranking program, ambient air pollution, land use, and global climate change were ranked highest as risks.

E2.7 RISK ASSESSMENTS IN THE REGULATORY PROCESS

We asked agencies how important risk assessments are in the regulatory process. Of the 117 agencies from the 42 states that engage in risk-related activities, 57 agencies in 32 states indicated that risk assessments are very important in the regulatory process. Twenty-nine agencies representing 20 states thought it was somewhat important, and 15 agencies in 14 states were unsure how important risk assessments are in the regulatory process. Only seven survey respondents from seven states indicated that risk assessments were not important at all in the regulatory process.

E2.8 DEGREE OF INFLUENCE RISK ASSESSMENTS HAVE ON AGENCY DECISION MAKERS AND LEGISLATORS

Our survey asked respondents to rate, with regard to risk the degree of influence (low, medium, or high) the following have on agency decision makers and legislators:

- 50 agencies in 35 states reported that quantitative health risk estimates have a high degree of influence on agency decision makers and legislators
- 32 agencies from 24 states reported court orders and 31 agencies representing 20 states reported budget and staff constraints as having a high degree of influence
- agencies indicated the influence of public opinion, interest groups, federal pressure, and influence of economic analyses all as having a moderate influence on agency decision makers and legislators
- ecologic risk estimates, environmental justice, and election pressure were reported as having a low degree of influence on agency decision makers and legislators

E2.9 PUBLIC INVOLVEMENT IN RISK ASSESSMENT

Very few state programs involve the public in all stages of risk assessments. In the scoping or early stages of the risk assessment process, 37 agencies in 26 states indicated that the public had no involvement. Eight agencies from seven states reported that the state selects individuals with whom to share information. Only four agencies from four states reported using open forums for sharing information and only four agencies indicated that in the scoping stages of risk assessment the public makes a recommendation that influences an agency decision. Only one agency reported that the public would make the decision regarding risk assessment activities. It is apparent that most state programs are targeted toward informing the public and not interacting with it in the early stages.

The survey asked the degree of public involvement in data collection activities regarding risk assessment. Thirty-five agencies in 25 states reported that the public has no involvement in data collection activities. Eleven agencies from 10 states reported that the state selects individuals with whom to share the information regarding data collection and very few, only six agencies in six states indicated they held open forums or that the public influences the outcome. No agencies indicated that the public makes the decision regarding data collection activities.

Thirty-six agencies in 26 states reported that the public had no involvement in data interpretation. Eleven agencies from eight states reported that the state selects individuals with whom to share the data-interpretation information, and 10 agencies in nine states said they hold open forums. Only three agencies from three states reported the public makes a recommendation which would influence the agency's decision regarding data interpretation.

Thirty agencies from 26 states reported the public had no involvement in risk characterization. Nineteen agencies from 15 states reported they held open forums for sharing risk characterization information. Ten agencies in eight states reported that the state selects individuals with whom to share information, and only four agencies from four states indicated that the public makes a recommendation that influences the agency's decision regarding risk characterization activities.

The survey results indicated that when agencies are publicizing the results of risk assessments, 32 respondents from 23 states said they held open forums to discuss the publication of risk assessment results and 20 agencies in 17 states reported that the public had no involvement in the publication of risk assessment results. Only six agencies from six states reported that the public makes recommendations which influence the decision, or that the state selects individuals with whom to share the risk assessment results.

E2.10 RISK MANAGEMENT

According to the National Academy of Sciences, risk management is: 'the evaluation of alternative risk control actions, selection among them including doing nothing, and their implementation' (National Research Council 1983, 1994). According to our survey results, of the 89 agencies that responded to the question, 76 agencies agree with the National Academy of Sciences' definition of risk management, and 13 used another definition.

Unlike risk assessment, risk management explicitly considers a broad range of legal, economic, political, and sociological factors. For analytical purposes risk assessment, risk management, and risk communication are often thought of as separate processes but in practice they frequently blend together.

Decision makers weigh several questions when considering the risks of a given situation. They consider who is responsible for the decision that is to be made and who is responsible for mitigating or compensating for damage. They need to consider the legal issues surrounding the alternatives and whether there are constraints on the possible options. Technical, biological, physical, and time limitations need to be considered. Decision makers also need to consider what resources are available for implementing the decision, i.e., what personnel and financial resources are available. To make a well-informed choice, decision makers require a wide range of knowledge. Risk management provides a context for balanced analysis and decision making.

The average number of full-time agency staff involved in risk management is 10. Most state officials involved in risk management are in water programs, followed by agriculture and occupational safety and health programs. There was roughly an equal distribution of staff involved in risk management among hazardous materials and solid waste programs, environmental and public health programs, and air programs. Natural resources programs had the fewest staff involved in risk management.

E2.11 ENVIRONMENTAL MANAGEMENT ACTIVITIES AND RISK ASSESSMENTS

Of environmental management activities where risk assessments are used they were reported most often in setting agency priorities and deciding permit applications. Hazardous materials and solid waste programs, environmental and public health programs, and air programs used risk assessments the most. Selecting targets for enforcement actions and long-term agency planning are areas that use risk assessments frequently in environmental management activities. This was followed by using risk assessments for public outreach and education. There was roughly an equal distribution of those who use risk assessments in negotiating with federal agencies and for short-term agency budget preparation.

E2.12 RISK COMMUNICATION

Subjective values will always influence the ranking of risks no matter how sophisticated the technical and analytical tools become. Communicating risk is not an easy task. Risk communication compresses detailed technical information, which can lead to the public's misunderstanding and distrust.

> 'Many people—including some scientists, decision makers, and members of the public—have unrealistic expectations about what can be accomplished by risk communication. For example, it is mistaken to expect improved risk communication to always reduce conflict and smooth risk management. Risk management decisions that benefit some citizens can harm others. In addition, people do not all share common interests and values, so better understanding may not lead to consensus about controversial issues or to uniform personal behavior. But even though risk communication cannot always be expected to improve a situation, poor risk communication will nearly always make it worse' (National Research Council 1989).

According to the National Academy of Sciences, risk communication is 'the exchange of information and opinion among individuals, groups, and institutions about the nature of risk or expressing concerns, opinions, or reactions to risk messages or to legal and institutional arrangements for risk management' (National Research Council 1983, 1994). The majority of state agencies surveyed agree with the National Academy of Sciences' definition of risk communication: 93 agencies reported they agree and six said they did not.

Our survey results indicated the average number of full-time agency staff committed to doing risk communication in an agency is four. Of that group, most are in occupational safety and health programs, followed by environmental planning programs. There was roughly an equal distribution of involvement for those in hazardous materials and solid waste, environmental and public health, water, and air programs. Agriculture and natural resource programs had the fewest people involved in risk communication activities.

E2.13 HOW RESULTS OF RISK ASSESSMENTS ARE COMMUNICATED TO THE PUBLIC

Most states believe that communicating risks to the public helps their agency during the regulatory process. For those involved in risk communication, the way they communicate results to the public most often is through public meetings and hearings. They also frequently communicate risk messages through written documents. Public availability sessions are used infrequently— 40 agencies from 24 states said they communicate in this way.

Very few agencies reported using outreach programs, and even fewer reported communicating results through other means.

E2.14 SPECIFIC PROGRAMS TO REPORT RISK ASSESSMENT

Our survey asked whether agencies have a specific program for including the public in risk assessments, reporting the results of risk assessments, or reporting the use of risk assessments in agency actions. Forty-nine agencies in 30 states indicated having a specific program for reporting the results of risk assessment in agency actions. Environmental and public health programs reported this most often. Thirty-three agencies from 21 states indicated they had a program for reporting the use of risk assessment, but more agencies and states, 36 and 24 respectively, reported they did not have a specific program. Thirty agencies in 22 states indicated they have a specific program for including the public in risk assessments, and again more agencies and states, 39 and 24 respectively, said they did not have a specific program.

E2.15 COMMUNICATING RISKS TO THE PUBLIC DURING THE REGULATORY PROCESS

Our survey asked how much communicating risk messages to the public helps an agency during the regulatory process. Agencies feel communicating risk to the public helped their agency during the regulatory process only somewhat; 45 agencies from 27 states indicated feeling this way. Hazardous materials and solid waste programs and environmental and public health programs reported this most, followed by air and water programs.

The survey results indicated 31 agencies in 21 states thought risk communication helped a great deal during the regulatory process. Nine agencies from eight states reported 'other', and only seven agencies in six states responded that risk communication did not help at all during the regulatory process.

E2.16 PUBLIC REACTION TO AGENCY'S RISK ASSESSMENT EFFORTS

Most state environmental agencies indicated it was difficult to gauge how the public generally reacted to their agency's efforts with respect to risk assessment. Fifty-three agencies in 29 states reported it is difficult to gauge; of that 53, hazardous materials and solid waste programs and environmental and public health programs reported this most often. Twenty-nine agencies from 20 states believed the public has reacted positively to their agency's efforts regarding risk assessment. Environmental and public health programs indicated this the most. Very few respondents, only seven agencies from seven states, indicated that the public has reacted negatively to their agency's efforts.

E2.17 CONCLUSIONS

This section illustrated that most states are actively engaged in risk-related activities. The bulk of these activities are focused on human health assessments in the area of hazardous and solid waste management. States also consider risk assessments to be a very important factor for developing regulations. Unfortunately, states have not developed adequate mechanisms to incorporate the public into this process. The majority of state risk communication activities occurs after an agency has made a decision not during the decision-making process. Although gauging public reaction to these outreach efforts is difficult, when an agency has been able to gauge the public's reaction it has been favorable.

E3 STATE USE OF RISK IN THE SITE REMEDIATION PROCESS

This section focuses on how states consider risk in the site remediation process. Specifically, how do they use risk to prioritize state cleanup decisions and set cleanup standards? The EPA has identified approximately 33 000 sites that may need to be addressed, only about 5% of which are serious enough to warrant cleanup under Superfund. Responsibility for ensuring that risks from the remaining sites are addressed falls upon the states. As states do not have the financial resources that the federal government does, examining state cleanup programs may reveal how states perceive the scientific application of risk-based environmental decision making.

In general, many states have found the EPA's process, which relies on a great deal of site-specific information and the use of site-specific risk assessments, problematic. Consequently, many states have developed their own risk-based procedures, which rely less on site-specific analysis. These approaches use risk in a more programmatic fashion and often do not fit the 'classical' definition of a site-specific risk assessment.

E3.1 WHY EXAMINE STATE USE OF RISK IN THE SITE REMEDIATION PROCESS?

As discussed in previous chapters, it is nearly impossible to develop a state perspective on risk. Not only do perspectives differ among states, there are often different viewpoints within states, between different regulatory programs, and within the same regulatory program. However, the previous section also demonstrated that risk is used pervasively in state environmental decision making. Therefore, to flesh out some of the issues raised in the first two sections, one regulatory program—site remediation—is analyzed to provide a more practical perspective on how risk is actually applied. Although air, drinking water, and occupational health and safety could have been examined, site remediation was chosen for several reasons:

- it illustrates how states use risk on different decision-making levels
- it has a large impact on state cleanup decisions at DOE facilities
- it is timely, as the EPA is developing risk-based soil screening (trigger) level guidance

By using risk both to set priorities and to set cleanup levels, site remediation can illustrate trends in how risk is applied on a variety of decision-making levels. For example, states use risk to prioritize cleanup initiatives by estimating the relative risks to human health and the environment posed by individual sites. States also set risk-based cleanup goals and cleanup criteria by defining acceptable levels of contamination. In both of these risk-based approaches, states are shying away from site-specific analyses that rely on enormous amounts of site-specific data. Instead, states are establishing cleanup priorities and developing cleanup goals using risk-based approaches that rely on average or otherwise hypothetical conditions.

Another reason for examining risk in the state site-remediation process is the impact state cleanup decisions may have on DOE facilities. Although the most contaminated DOE sites are on the National Priorities List (NPL), state remediation programs play an important part in the cleanup process. The implementation of the Federal Facilities Compliance Act also gives states a larger part in regulating DOE facilities. Understanding how states incorporate risk-based decision making into their cleanup programs could promote smoother relations among state and federal agencies.

Site remediation also helps to explain how states incorporate risk in environmental decision making because the EPA is developing 'soil trigger levels' for a variety of contaminants. This effort is a result of EPA's 30-day study to outline options for accelerating the rate of cleanups at Superfund sites. Development of soil trigger levels was one element of the suggested improvements. A trigger level or screening level is a chemical concentration in soil below which there is not sufficient concern to warrant further site-specific study. Although state policy is often based on federal regulations and statutes, EPA policy is often driven by state innovations, and therefore it is important to understand how states incorporate such standards into their own programs.

E3.2 STATE EXPERIENCES WITH RISK IN THE SUPERFUND PROCESS

Most of the national attention focused on the cleanup of hazardous waste sites revolves around the NPL. Sites on the NPL have scored high enough by the EPA's ranking process to become eligible for Superfund resources. However, the approximately 1250 sites on the NPL represent only a small fraction of the total number of contaminated sites that may require remediation. For example, only about 5% of the total sites that are in the EPA's Comprehensive Environmental Response, Compensation and Liability Act (CERCLA) Information System (CERCLIS) are included on the NPL. The rest are the responsibility of the states to determine whether and to what extent cleanups may be required. Therefore, how states incorporate the risk provisions of Superfund provide insights into how states use risk in their own programs.

Trends indicate that states are using risk-or health-based standards more and more in the site remediation process. In 1991 38 states considered some form of risk analysis to determine cleanup standards, an increase of six from the previous year (Environmental Law Institute 1992). By 1994 this number could exceed 45. These states either set predetermined concentration levels for cleanup or conduct site-by-site risk assessments. State applications of both approaches differ greatly.

States that set risk-based concentration levels apply them either in specific situations or generally to a broad range of sites. Similar to the relationship between 'applicable or relevant and appropriate requirements' (ARARs) and risk-based cleanup levels under Superfund, some states only rely on risk-based cleanup levels in the absence of other existing standards. (ARARs are other requirements that Superfund sites may have to comply with in addition to CERCLA requirements.) Indiana, for example, uses risk-based cleanup levels only when a maximum contaminant level (MCL) for drinking water does not exist or where multiple carcinogens are present. Other states, such as Oregon and Texas, apply cleanup standards more generally.

States that rely on site-specific risk assessments to set cleanup standards generally follow the EPA's guidelines; however, how such assessments are used varies. Massachusetts, for example, only conducts a site-specific risk assessment when the site conditions deviate dramatically from the average conditions that were used to develop general standards. Kentucky and Florida, in contrast, weigh the results of site-specific risk assessments with other applicable standards to determine final cleanup levels (Environmental Law Institute 1992).

Although states are incorporating health-based criteria into site cleanup decisions, trends indicate that states are steering away from site-specific risk assessments in favor of risk-based concentration levels that apply to a majority of state sites. One reason for this shift is the states' experience with Superfund's baseline risk assessment process. States find the EPA's baseline risk process problematic for several reasons:

- it is too ambiguous as it uses a risk range (10^{-4}–10^{-6}) that leads to inconsistent cleanup decisions
- it requires too much site-specific data and analysis
- it diverts resources away from actual cleanup activities
- it is not incorporated into the entire remediation process
- the results are very difficult to defend because they are so contentious and uncertain

E3.3 STATE CLEANUP LEVELS

To address these issues many states have developed generic cleanup standards. (Cleanup standards are actual levels that remediation must achieve as opposed to screening levels which are used to determine if further action is required.) Such programs identify the contaminant levels that remedial actions must achieve with less need for site-specific data than the traditional site-specific (baseline) risk assessments required under Superfund's National Contingency Plan. Under the EPA's program, cleanup officials must treat each site uniquely, and extensive risk-based data must be collected at every site to determine allowable cleanup levels. The use of cleanup standards, however, reduces data requirements by indicating the appropriate contaminant levels for all sites or various classes of sites. Except where officials identify unique site conditions that may warrant a modification of the standards, this approach avoids the process of setting cleanup levels at every site.

The use of cleanup standards is based on the presumption that certain contaminant levels will achieve acceptable risk targets at almost all sites. To set the standards, officials estimate the risk reduction achieved by cleanup levels at previous (as well as computer-simulated) remedial actions. Results that consistently achieve desired risk-reduction targets are then generalized into cleanup standards that apply to all sites or to a large segment of the total site universe as the cleanup levels a remedial action must achieve, without collecting extensive data at a specific site. In addition to reducing the reliance on baseline risk assessments, states also have found that using cleanup standards eliminates negotiations among 'potentially responsible parties' (PRPs) and cleanup program officials at each site over how extensive a cleanup should be.

E3.4 CLEANUP LEVEL CHARACTERISTICS

Currently, 10 states have been identified that use or are about to promulgate soil cleanup standards (see Table E1). These 10 states include Arizona, Massachusetts, Michigan, Minnesota, New York New Jersey, Oregon, Texas, Washington, and Wisconsin. This represents an increase of 50% since 1989. The number is increasing rapidly, as 26 states are considering or planning on issuing cleanup standards over the next few years (US EPA 1993).

Generally, the majority of state cleanup levels are implemented through state regulations rather than guidance or agency policy. Although all state regulations are based on statutory authority, the specificity of state laws vary. Some state regulations are a direct response to a statutory mandate. Washington, for example, developed its regulations in response to a statute that specified the exact nature of the cleanup levels. Other states, such as Texas, developed their regulations based on the agency's desire to respond better to the regulatory community. In this case, Texas' regulatory authority is based on an existing statutory authority that is very broad. In other states, the legislature may require the state to develop cleanup standards without mandating any specific requirements.

Although the majority of standards are contained in state regulations, some state cleanup levels are contained in guidance. Generally in these cases, the states use specific cleanup levels to aid the cleanup process, not as the main

Table E1 Profile of state cleanup program

State	Industrial and residential scenarios	Different aquifer classifications	Site-specific conditions
Arizona	Yes	Yes	Yes
Massachusetts	Yes	Yes	Yes
Michigan	Yes	Yes	Yes
Minnesota	No	No	Yes
New Jersey[a]	Yes	Yes	No
Oregon	Yes	No	Yes
Texas	Yes	Yes	No
Washington	Yes	No	No
Wisconsin	Yes	No	No
New York	No	Yes	Yes

[a]New Jersey's standards are currently in draft form.
Source: National Governor's Association from phone interviews, review of state documents, and EPA survey.

driving factor. New York and Minnesota, for example, use generic cleanup standards in their preliminary assessment program to determine whether a cleanup is required.

Most states that have developed cleanup standards have done so for soil. The number of contaminants for which states have developed soil standards varies. Michigan, for example, has developed standards for 200 contaminants, while Washington has developed standards for only 20. In developing specific contaminant levels the majority of states use a specific risk target for carcinogens of 10^{-6}. Massachusetts, for example, has developed a residential soil cleanup level for arsenic of 20 parts per million (p.p.m.). In developing this standard the state used various exposure and dose–response assumptions to calculate that 20 p.p.m. presents a 1 in 1 000 000 chance of developing cancer. Some states use less-stringent standards for noncarcinogenic compounds. New York for example, uses a 10^{-5} risk target for calculating class C (noncarcinogenic) compounds.

States use different exposure equations to calculate target values when developing risk-based concentration levels for soil. For example, Arizona uses an exposure frequency of 365 days for children under residential scenarios. Massachusetts uses an exposure frequency of 153 days for the same scenario. Exposure frequencies measure how long a person is exposed to a contaminant. The results are fed into a model that calculates the final concentration level. States also use different exposure pathways to calculate cleanup levels. Although all states have identified soil ingestion as one exposure pathway, many states use various combinations of other pathways to calculate standards. For example, Wisconsin uses inhalation of dust as a second pathway, while Michigan uses dermal contact.

Another important factor states use in developing soil cleanup levels are land-use criteria. Many states have developed separate values for residential and industrial levels. Only Minnesota and New York were identified as lacking specific provisions for differentiating between residential and industrial land use. The majority of states that do differentiate between land uses calculate less-stringent standards by changing the exposure assumptions, not by modifying the risk target of 10^{-6}. However, some states, such as Washington, do use a 10^{-5} standard in calculating industrial contaminant levels (US EPA 1993).

E3.5 STATE CLEANUP LEVEL APPROACHES

State approaches to setting cleanup standards generally fall into three categories: a single set of uniform numeric standards; a series of options from which to choose standards; and streamlined procedures for setting site-specific cleanup levels.

New Jersey's 1992 proposed cleanup standards exemplify the first approach of establishing uniform numeric cleanup levels. State regulators based the standards primarily on human health criteria designed to achieve a 'one-in-a-million lifetime cancer risk level'. All remedial activities performed under any program area are subject to the standards, including Superfund, non-NPL, RCRA corrective actions, and voluntary cleanups. The cleanup standards apply to any site with contaminants in concentrations higher than the concentrations set in the standards. The standards apply to preexisting rather than future hazardous waste releases.

New Jersey's proposed cleanup standards contain numeric levels both for soils and groundwater. Soil standards are broken into two subgroups. Surface soil standards account for contact between a human receptor and contaminated soil, and subsurface soil standards address the transfer of contaminants from soil to groundwater and surface water. A PRP may choose higher (less-protective) soil standards if the PRP can limit the site to nonresidential activities. Groundwater standards are broken into subclasses corresponding to the existing or planned quality/use classifications that the state has assigned to its groundwater resources. Procedures allow the department to set more stringent standards to prevent ecological impact or to establish additional standards at contaminated sites with substances not listed in the uniform standards. New Jersey's proposal also would allow PRPs to petition for an alternative cleanup standard or deferral of a standard. The standards also include mechanisms to verify compliance with both soil and groundwater cleanup levels.

Washington State's cleanup standards represent the second approach of a series of options for setting cleanup levels that account for the different conditions encountered at contaminated sites. The state's first option requires compliance with standards already established for 25 of the most common hazardous substances found at contaminated sites. The option is for cleanups that are relatively straightforward or involve only a few contaminants, each of which must be listed in the state standards. This approach best serves small sites that do not warrant the cost of conducting risk assessments and site studies. The second option is similar to that used under Superfund for setting cleanup levels, i.e., using site-specific risk assessments. Unlike Superfund, however, this option uses a single risk target instead of a risk range. The risk level for individual carcinogens cannot exceed 1 in 1 000 000 while the total risk from multiple pathways cannot exceed 1 in 1000. Cleanup officials may use the third option when levels set under the first two options are technically impossible to achieve, are lower than background concentrations, or may cause more harm than good (e.g., when achieving the more stringent levels would require destroying important wildlife habitat). This approach sets a lifetime cancer risk of 1 in 100 000 for both individual substances and the total risk caused by all substances on the site.

The Texas Water Commission recently passed regulations that provide options for setting cleanup standards. Texas' regulations offer PRPs three choices, each involving different cleanup standards and oversight requirements. The first option requires cleanup to background levels. In return for achieving this most-stringent cleanup level, the state does not require PRPs to perform post-closure care at the site. The second option requires the PRP to meet cleanup standards designed to achieve a 10^{-6} risk target. The third option would not use cleanup standards, in favor of the CERCLA process of setting cleanup levels based on site-specific risk assessments. The state designed this three-tiered approach to allow the state and PRPs to use standard setting processes that are appropriate to the severity of a particular contaminated site. This allows the state to vary data collection, oversight, and post-closure care requirements according to the stringency of each option.

Illinois uses the third approach for setting cleanup standards—a streamlined process that establishes site-specific cleanup levels. The Illinois approach applies to all remedial activities, including NPL sites, RCRA closures, RCRA corrective actions, voluntary cleanups, state-conducted removals, and underground storage tank cleanups. The approach relies on a two-stage process conducted by the Illinois Environmental Protection Agency (IEPA) to establish cleanup levels for a specific site.

Initially, the state remedial project manager (RPM) prepares a request for cleanup objectives in a standardized format and submits the request to the agency's staff toxicologists to evaluate risk and to promulgate cleanup objective concentrations for various environmental media and routes of exposure. The IEPA toxicologists' determinations are based on protection of human health and the environment and do not consider costs.

The RPM may concur with the health-based cleanup objectives or may appeal the determination to the Cleanup Objectives Review and Evaluation (CORE) group, a body of higher-level managers who may consider costs and other relevant factors. The CORE group may accept or modify the health-based cleanup objectives or may approve alternative approaches to treatment of the waste.

E3.6 IMPLEMENTATION ISSUES

As state programs for developing cleanup levels are relatively new, states have not fully evaluated how well these programs are working. However, initial findings indicate that cleanup standards generally are being well received by industry and the public. States have noted, however, a number of concerns that have arisen during the implementation of such standards. A major concern is flexibility. Industry has raised a number of concerns over the lack of flexibility in the use of cleanup standards, pointing to difficulties of applying standards based on general conditions to unique circumstances surrounding an individual site. New Jersey, for example, received approximately 4000 comments on its proposed soil standards regulations. Many of these comments were requests from industry to provide a mechanism for addressing site-specific standards. New Jersey is in the process of evaluating these comments.

To accommodate concerns over flexibility, many states allow responsible parties to modify standards if they can prove that site characteristics differ significantly from the assumptions used to develop generic cleanup standards. Washington State, for example, has developed cleanup-level ranges if the responsible party can prove that the site deviates from the generic model. Other states, such as Massachusetts and Texas, have developed approaches that give responsible parties the choice of conducting a baseline risk assessment if they do not believe the cleanup standards are applicable.

E3.7 CONCLUSIONS

Many states use risk-based standards to set site remediation cleanup levels; however, most states do not favor Superfund's approach, which calls for site-specific risk assessments using a risk range of 10^{-4}–10^{-6}. In order to streamline and standardize the process states have developed generic cleanup levels based on a specific risk target, usually 10^{-6}. To calculate the toxicity of a given exposure to a particular contaminant the majority of states use the EPA's standard risk equations, but modify the exposure pathways and assumptions to determine the specific contamination levels that a site must achieve.

States generally believe that cleanup standards can simplify, standardize, and streamline the remediation process and eliminate much of the contention over 'how clean is clean', thereby speeding up remediation. Cleanup standards also reduce reliance on a baseline risk assessment, which some states find problematic due to the ambiguity and reliance on site-specific data. States also believe that cleanup standards will encourage voluntary cleanups because all parties involved will have a clear idea of the specific cleanup goals to be met.

States realize that applying general cleanup standards to all sites is difficult. Consequently, some states have developed separate standards for different scenarios, such as residential and industrial land uses. Other states give responsible parties the option of conducting a baseline risk assessment if they believe the standards are not applicable.

E4 PRIORITIZING SITE CLEANUPS

States also use risk to prioritize site-remediation activities. The federal Superfund program does not have a national approach to prioritizing sites once they are listed on the NPL. Many states, however, have developed mechanisms to prioritize cleanup decisions to ensure that sites posing the most risk to human health and the environment are addressed first. The states' interest in prioritization is largely a response to resource constraints and the large number of sites that must be remediated. The New Mexico cleanup program for example, has an annual budget of approximately $200 000 and 10 staff devoted to cleanup of over 100 sites. If the state was to address all sites simultaneously, this resource level allows one state official to oversee 10 sites with a budget of only $2000 per site. Consequently, the state developed a system to ensure allocation of resources to sites that posed the most risk.

States also have found that using systematic approaches to prioritizing sites not only helps allocate resources, but also helps defend state cleanup decisions to the general public and industry. Washington and Oregon officials discovered that when they began implementing their program they were able to communicate cleanup decisions more effectively to the state legislature, industry, and the public. Prior to the development of their ranking processes both states spent a considerable time justifying cleanup decisions by explaining why one particular site was being addressed instead of another.

States' approaches to prioritizing site cleanups generally fall into two categories: quantitative or qualitative approaches (see Table E2). A quantitative approach involves assigning a numerical rank to each site to measure the relative risk a site poses compared with other sites. Such approaches are based on the same principle as EPA's Hazard Ranking System (HRS). HRS assigns a score to each site based on an evaluation of exposure pathways and the types and amounts of contaminants that are present at the site. Under the HRS most state programs similar to the HRS are used specifically to establish the cleanup

Table E2 Overview of state site priority approaches

EPA region		State HRS-I	EPA HRS-II	Quantitative	State developed model Qualitative[a]
1	Connecticut			X	
1	Massachusetts			X	
1	Maine				X
2	New Jersey			X	
2	New York				X
3	Pennsylvania		X		
4	Florida				X
4	Mississippi	X			
5	Illinois				X
5	Michigan			X	
5	Minnesota	X			
5	Ohio				X
5	Wisconsin	X			
6	New Mexico			X	
6	Texas		X		
8	Colorado				X
9	California	X			
10	Oregon			X	
10	Washington			X	

[a]Qualitative approaches do not numerically rank each site. Cleanup decisions are based on broad-based criteria or best professional judgement by agency staff.
Source: National Governors' Association from phone interviews and review of state documents.

program's priorities. In contrast, qualitative approaches for prioritizing sites do not assign specific scores for each site. Instead, the relative risk posed by sites is measured using broad categories (such as high, medium, and low) or best professional judgement.

ACKNOWLEDGMENTS

The editor gratefully acknowledges permission to excerpt material in this chapter from Risk in Environmental Decision Making. A State Perspective. Working Paper prepared by the National Governors' Association, 444 North Capitol Street, Washington, D.C. 2001, USA.

REFERENCES

Environmental Law Institute (1992). *An Analysis of State Superfund Programs: 50 State Study.* 1616 P St., NW Washington, DC 20036.
National Research Council (1983). *Risk Assessment in the Federal Government: Managing the Process.* National Academy Press, Washington, DC.
National Research Council (1989). *Improving Risk Communication.* Committee on Risk Perception and Communication, National Academy Press, Washington, DC.
National Research Council (1994). *Science and Judgement in Risk Assessment.* National Academy Press, Washington, DC.
US Environmental Protection Agency (1993). *EPA Office of Superfund Revitalization, State Soil Cleanup Level Survey.* Washington, DC.

Appendix F

Water Pollution: Differences among the States in Issuing Permits Limiting the Discharge of Pollutants

UNITED STATES GENERAL ACCOUNTING OFFICE
GAO/RCED-96-42 January 23, 1996

F1 BACKGROUND

The objective of the Clean Water Act was to restore and maintain the chemical, physical, and biological integrity of the nation's waters. The Congress established a series of national goals and policies to achieve this objective, including what is referred to as 'fishable/swimmable' water quality. That is, whenever attainable, the quality of the water should be such that it provides for the protection of fish, shellfish, wildlife, and recreation in and on the water. To help meet national water quality goals, the act established the National Pollutant Discharge Elimination System (NPDES) program, which limits the discharge of pollutants through two basic approaches: one based on technology and the other on water quality.

Under the technology-based approach, facilities must stay within the discharge limits attainable under current technologies for treating water pollution. The Enviromental Protection Agency (EPA) has issued national minimum technology requirements for municipal facilities and 50 categories of industrial dischargers. The states' and the EPA's permitting authorities use these requirements to establish discharge limits for specific pollutants.[1] In contrast, under the water-quality-based approach, facilities must meet discharge limits derived from states' water quality standards, which generally consist of: (1) 'designated uses' for the water bodies (e.g., propagation of fish and wildlife, drinking water, and recreation), and (2) narrative or numeric criteria to protect the designated uses.[2] Narrative criteria are generally statements that describe the desired water quality goal, such as 'no toxics in toxic amounts'. Numeric

---

[1]If technology standards do not exist for a particular industry, the permitting authorities consider other pertinent data and use their 'best professional judgement' to establish the discharge limits.

[2]In addition, the states are required to include an 'antidegradation' provision in their water quality standards to ensure that in those water bodies where the quality exceeds that required to support the designated use, the quality in general will be maintained at the existing level.

This appendix is a US Government work and, as such, is in the public domain in the United States of America.

criteria for specific pollutants are generally expressed as concentration levels and target certain toxic pollutants that the EPA has designated as 'priority pollutants'.[3]

In addition to adopting water quality standards, the states may also establish policies concerning certain technical factors that affect the implementation of the standards in the discharge permits. For example, many states have adopted policies for: (1) establishing mixing zones (limited areas where discharges mix with receiving waters and where the numeric criteria can be exceeded); (2) determining the amount of available dilution (the ratio of the low flow of the receiving waters to the flow of the discharge); and (3) considering background concentration (the levels of pollutants already present in the receiving waters).

When the states' and EPA's permitting authorities are deciding how extensively the pollutants should be controlled in a facility's permit, they first look to the technology-based standards. If the discharge limits derived by applying these standards are not low enough to protect the designated uses of the applicable water body, the permitting authorities turn to the state's water quality standards to develop more stringent limits. To achieve the tighter limits, a facility may have to install more advanced treatment technology or take measures to reduce the amounts of pollutants needing treatment.

This report focuses on the water-quality-based approach to controlling pollution and the way the states' and EPA's permitting authorities are implementing water quality standards in the NPDES permits issued to 'major' facilities.[4] As of July 1995, approximately 59 000 municipal and industrial facilities nationwide had received permits under the NPDES program, and about 6800 of these permits were for major facilities, including about 4000 municipal facilities and 2800 industrial dischargers.

F2 THE STATE'S AND THE ENVIRONMENTAL PROTECTION AGENCY'S PERMITTING AUTHORITIES VARY IN WHETHER AND HOW THEY CONTROL POLLUTANTS IN DISCHARGE PERMITS

Our review of the data on municipal permits for five commonly discharged toxic pollutants disclosed that decisions about whether and how to control

---

[3]As a result of a 1976 consent decree, the EPA was required to publish water quality criteria to protect human health and/or aquatic life for a specified set of pollutants or classes of pollutants by 1979. The Congress later specifically designated these same chemicals as toxic pollutants under section 307(a) of the Clean Water Act, and the EPA selected 126 key chemicals or classes within this group for priority status. For other, nonpriority pollutants, the states may choose to adopt either numeric or narrative criteria.

[4]The EPA classifies certain facilities as major, usually on the basis of their capacity and/or the type and quantity of pollutants they discharge.

pollutants differed both from state to state and within states. In some instances, differences in the limits themselves, or in the standards and policies used to derive the limits, have led to concerns between neighboring states.

F2.1 SOME PERMITTING AUTHORITIES IMPOSE DISCHARGE LIMITS, OTHERS REQUIRE MONITORING ONLY, AND SOME IMPOSE NO CONTROLS

Using the EPA's Permits Compliance System database, we extracted data on the 1407 permits issued to municipal wastewater treatment facilities nationwide between February 5, 1993, and March 21, 1995,[5] to determine what types of controls, if any, the states' and EPA's permitting authorities had imposed in these facilities' discharge permits for five toxic metal pollutants: cadmium, copper, lead, mercury, and zinc.[6] We found that when the permitting authorities decided that some type of control was warranted, some consistently established numeric discharge limits in their permits, and others imposed monitoring requirements in all or almost all cases.

For example, North Carolina issued 93 permits during our review period and, whenever it determined that a pollutant warranted controls, it always established numeric discharge limits rather than impose monitoring requirements. Other states, such as New York and West Virginia, also consistently established numeric limits for controlling the five pollutants we examined. In contrast, New Jersey issued 44 permits during our review period and, except for one permit that contained a limit for copper, the state always imposed monitoring requirements instead of discharge limits when the state determined that controls were warranted. Oregon, among other states, made similar decisions, as did EPA's Region VI when it wrote permits for Louisiana, a state not authorized to issue NPDES permits.

We also found that some states, such as Vermont and Arkansas, had not imposed discharge limits or monitoring requirements in the following instances in which the EPA's regional officials said that such controls may be warranted.

1. In Vermont, none of the discharge permits for major municipal facilities contained discharge limits or monitoring requirements for the five metals. However, at our request, the cognizant EPA regional staff in Region I reviewed four of the 15 municipal permits issued by Vermont and determined that for

---

[5]We chose this time frame because as of February 5, 1993, all of the states had water quality standards for toxic pollutants and were required to use these standards to derive discharge limits for all new permits and permits up for renewal.

[6]We selected these five pollutants for our review because, according to officials in the EPA's Permits Division, they are commonly discharged by municipal wastewater treatment facilities and are likely to be found nationwide.

two of the facilities, limits or monitoring requirements would probably be appropriate. Vermont officials agreed to review the permits and consider additional requirements.

2. Arkansas, with one exception, had not imposed either limits or monitoring requirements in its municipal permits for the toxic metals we examined. State officials are allowing these facilities to continue operating under 'old' permits rather than reissuing them. The officials told us that if the permits were to be formally reopened, the state would be obligated to apply the EPA-imposed water quality standards for the metals. Arkansas officials believe that these standards are too stringent and that the facilities would engage the state in a costly appeal process if limits were imposed. Officials from the cognizant EPA regional office (Region VI) said that the Arkansas permits should contain discharge limits but that the EPA does not have the authority to impose such limits in a state authorized to issue permits when the state simply declines to reissue them. The EPA's only recourse would be to take back responsibility for the program—an unrealistic option.

For facilities, both monitoring requirements and discharge limits can be costly to implement. According to officials in the EPA's Permits Division, the costs of monitoring depend on the frequency of required sampling and on the types of pollutants that must be tested. The costs of installing advanced treatment equipment to meet discharge limits are usually much higher. These officials also said that because of these differences in cost, the facilities that are subject to monitoring requirements generally enjoy an economic advantage over the facilities that must meet discharge limits, all other things being equal. Furthermore, the facilities that are subject to neither type of control enjoy an economic advantage over the facilities that must comply with limits or monitoring requirements.

Overall, our analysis disclosed that for each of the five pollutants, about 30% of the permits contained limits or monitoring requirements, while about 70% contained neither type of control. According to the EPA's permitting regulations and guidance, there can be legitimate reasons for imposing no controls over some pollutants: Generally, either the facilities are not discharging the pollutants or their discharges are deemed too low to interfere with the designated uses of the applicable water bodies.

F2.2 DISCHARGE PERMITS MAY CONTAIN DIFFERENT LIMITS FOR THE SAME POLLUTANTS

The EPA and the states agree that differences in the numeric limits for specific pollutants can and do exist—not only from state to state, but from water body to water body. To illustrate these differences, we extracted data on numeric limits as part of our analysis of the EPA's data on municipal permits. Specifically, from the 1407 permits for municipal facilities issued nationwide between

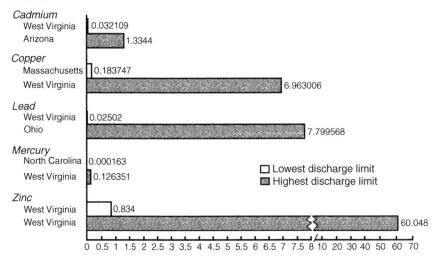

Figure F1 Range of differences in discharge limits for five toxic metal pollutants at facilities discharging into freshwater. For each of the five pollutants a range for the facilities' capacity was selected that (1) was sufficiently narrow for the facilities contained within it to be considered to be similar in size (eg. a range of 1.4–2.5 million gallons per day) and (2) included as many permits as possible for our analysis. For the sake of consistency, we used the 'daily maximum' discharge limits for our analysis. Some permits also contained other types of limits, such as limits on the weekly or 30-day average discharge. After applying the selection criteria we were left from 19 to 34 permits issued nationwide for each of the five pollutants. The discharge limits were expressed as maximum concentrations — either as milligrams per liter or micrograms per liter. For the sake of consistency, we converted all limits to pounds per day.

February 5, 1993, and March 21, 1995, we identified those facilities discharging into freshwater:[7] (1) whose permits contained discharge limits for one or more of the five toxic metals, and (2) whose plant capacity, or design flow,[8] was included in the EPA's database.

For each of the five pollutants, we found significant differences in the amounts that facilities were allowed to discharge across the nation—even for facilities of similar capacity. In the case of zinc, both the highest and the lowest limits were established in the same state. Figure F1 shows the results of our analysis.

As Figure F1 indicates, differences in the numeric limits for the same pollutant can be significant—in the case of mercury, about 775 times greater at

---

[7]We excluded permits for facilities that discharge into marine waters because the limits derived for these facilities can be significantly higher than the limits established for facilities that discharge into freshwater. Including such facilities would, therefore, have distorted our analysis

[8]A facility's capacity is stated in terms of average design flow, or the amount of wastewater that the facility is designed to discharge, in millions of gallons per day.

one facility than at another facility of similar capacity. We discuss the causes of the differences in discharge limits later in this report.

F2.3 VARIATIONS IN STATES' NPDES PROGRAMS HAVE RAISED CONCERNS IN NEIGHBORING JURISDICTIONS

Variations in the discharge limits, or in the standards and procedures used to derive these limits, have been a source of concern, particularly when neighboring jurisdictions share water bodies and the differences are readily apparent to the permitting authorities and discharging facilities, as the following examples illustrate:

1. In 1995, an industrial facility in Pennsylvania challenged a discharge limit for arsenic because Pennsylvania's numeric criterion was 2500 times more stringent than that used by the neighboring state of New York, into which the discharge flowed.[9] Among other things, the discharger argued that having to comply with the more stringent criterion created an economic disadvantage for the company. Eventually, Pennsylvania agreed to reissue the permit with a monitoring requirement for arsenic instead of a discharge limit. The state has also revised its water quality standards using the less stringent criterion.

2. Oklahoma challenged the 1985 permit that EPA issued to an Arkansas municipal wastewater treatment facility that discharges into a tributary of the Illinois River. One of the key issues in the case was Oklahoma's contention that the facility's permit, which was based on Arkansas's water quality standards, contained limits that would violate Oklahoma's water quality standards when the facility's discharge moved downstream. As a result, Oklahoma officials maintained, the river would not achieve its designation as 'outstanding natural resource water', a special classification designed to protect high-quality waters. Although the EPA has the authority to ensure that discharges in the states located upstream do not violate the water quality standards in the states located downstream, the agency determined that this case did not warrant such action, in part because the discharge allowed under the permit would not produce a detectable violation of Oklahoma's standards. In 1992, the Supreme Court ruled that the EPA's issuance of the Arkansas permit was reasonable.[10]

Concerns among states about differences in water quality standards and the policies that affect their implementation may become more common in the

---

[9]Pennsylvania had updated its water quality standards on the basis of current information on health effects published by the EPA; the state adopted a standard of 0.02 μg/L for arsenic. New York continued to rely on the EPA's earlier guidance on arsenic and maintained its standard at 50 μg/L.
[10]Arkansas v. Oklahoma, 503 U.S. 91 (1992).

future. According to a recent analysis by the US Geological Survey,[11] many states receive more than half of their water pollution from neighboring states. While much of this pollution may be attributed to diffuse—or 'nonpoint'—sources, such as agricultural runoff, according to an official from the US Geological Survey, the discharges from municipal and industrial facilities allowed under permits also contribute to interstate pollution.

F2.4 SEVERAL FACTORS CONTRIBUTE TO DIFFERENCES IN CONTROLLING POLLUTANT DISCHARGES

Both the act and the EPA's regulations give the states and the EPA considerable flexibility in implementing the NPDES program. The permitting authorities differ considerably in how they assess the likelihood that the states' water quality standards will be exceeded, as well as in how they decide what controls are warranted. If they decide that discharge limits are warranted, these limits can differ widely because of differences in the (1) states' water quality standards and (2) implementation policies that come into play when the permitting authorities 'translate' general water quality standards into limits for specific facilities in specific locations.

F2.5 PERMITTING AUTHORITIES DIFFER IN HOW THEY DECIDE ON THE NEED FOR CONTROLS OVER POLLUTANT DISCHARGES

We found differences in how the permitting authorities determine that a pollutant has the 'reasonable potential' to violate a state's water quality standard and prevent the designated use of a water body from being achieved. In EPA's Region I, for example, the permitting officials believe that one or two samples indicating the potential for a violation may suffice to justify imposing a discharge limit. In contrast, given the same evidence, officials in EPA's Region VI generally impose requirements for monitoring in order to collect data over a longer period of time—up to the 5-year life of the permit.

Officials in the Permits Division at the EPA headquarters agreed that there are differences in how the states' and the EPA's permitting authorities decide whether and how to impose controls over pollutant discharges. The officials said that a key element in these differences is the amount and type of data the authorities require to determine reasonable potential; some permitting authorities are comfortable with establishing discharge limits on the basis of limited information, while others want to collect more data and impose monitoring requirements. To assist the states and EPA's regional offices, EPA

---

[11]The US Geological Survey issued an abstract of its findings in the spring of 1995 and plans to publish a full report by the spring of 1996.

APPENDIX F

has issued national guidance,[12] including a suggested methodology and other options for determining reasonable potential. However, Permits Division officials emphasized that the law and applicable regulations provide for flexibility in decisions on reasonable potential and other aspects of the NPDES program.

F2.6 SETTING DISCHARGE LIMITS OFFERS MANY OPPORTUNITIES FOR PERMITTING AUTHORITIES TO EXERCISE FLEXIBILITY

The states have exercised the flexibility available within the Clean Water Act and EPA's regulations to: (1) adopt different water quality standards, and (2) apply different policies in implementing these standards in permits. As a result of these differences, discharge limits can vary significantly even, as illustrated earlier, for facilities of similar capacity.

F2.6.1 Differences in water quality standards

In the case of the states' water quality standards, the designated use assigned to a particular water body can affect how stringent a facility's discharge limit will be. For example, if a facility is discharging into a water body designated for recreational use, the discharge limits are likely to be less stringent than they would be if the water body was designated for use as a drinking water supply.

Water quality standards also differ in terms of the numeric criteria the states adopt to ensure that the designated uses of the water will be achieved or maintained. The EPA has provided guidance to the states on developing these criteria.[13] Some states have adopted the EPA's numeric criteria (e.g., a human health criterion for mercury that allows for no more than 0.144 µg/L) as their own, and others have developed different criteria that reflect regional conditions and concerns. For example, Texas modified the EPA's criteria to account for higher rates of fish consumption in the state.

Another significant source of differences in the states' water quality standards is the cancer risk level that is selected for carcinogenic pollutants.[14]

---

[12]US Environmental Protection Agency (1991). *Technical Support Document for Water Quality-based Toxics Control.* Office of Water, Washington, DC.

[13]Most recently, the EPA has issued new guidance on establishing numeric criteria for metals. The states will be allowed to use either of two methodologies. As a result, some states will have more stringent criteria for toxic metals than others.

[14]The risk level, in this context, is the probability of additional cancer cases in a population as a result of exposure to toxic pollutants. All other things being equal, a water quality standard based on a risk level of 1 additional cancer case per 100 000 people is less stringent than a standard based on a risk level of 1 additional cancer case per 100 000 people. The EPA issues criteria for protecting human health using risk levels ranging from 1 additional cancer case per 100 000 people to 1 additional case per 1 000 000 people, and the states have the discretion to base their own standards on any risk level within this range. The states may use other risk levels if such levels are scientifically defensible.

For example, Connecticut typically bases its numeric criteria for these pollutants on a risk level of 1 excess cancer case per 1 000 000 people, while Arkansas bases its criteria on a risk level of 1 excess cancer case per 100 000 people. Thus, Connecticut's criteria are 10 times more stringent than Arkansas's.

F2.6.2 Differences in implementation policies

Many states have established implementation policies that can significantly affect the application of water quality standards in establishing the discharge limits for individual facilities. These policies address technical factors such as mixing zones, dilution, and background concentration.

The states differ in their policies for mixing zones: limited areas where the facilities' discharges mix with the receiving waters and numeric criteria can be exceeded. The states' policies can influence the stringency of the discharge limits by restricting where such zones are allowed and/or by defining their size and shape. In Texas, for instance, the size of mixing zones in streams is typically limited to an area 100 feet upstream and 300 feet downstream from the discharge point; other states apply different standards or do not allow mixing zones in some types of water bodies. In general, the discharge limits will be less stringent for a facility located in a state that allows mixing zones than for a facility in a state that requires facilities to meet numeric criteria at the end of the discharge pipe.

The states' policies on dilution—the ratio of the low flow of the receiving waters to the flow of the discharge—can also influence the stringency of the discharge limits. In general, the larger the volume of the receiving waters available to dilute, or reduce the concentration of, the pollutants being discharged, the less stringent the discharge limit. Thus, all other things being equal, the discharge limit for a facility located on the Mississippi River will be less stringent than the limit for a similar facility located on a smaller river. The states also use different assumptions in computing the flow of a facility's discharge (e.g., the highest monthly average during the preceding 2 years or the highest 30-day average expected during the life of the permit) and the low flow of the receiving waters (e.g., the lowest average flow during 7 consecutive days within the past 10 years or the lowest 1-day flow that occurs within 3 years).

The states also have different policies on background concentration—the level of pollutants already present in the receiving waters as a result of naturally occurring pollutants, permitted discharges from upstream, spills, unregulated discharges, or some combination of these sources. In general, the higher the level of the background concentration, the more stringent the discharge limit will be because the extent of the existing pollution affects the amounts that facilities may discharge without violating the water quality standards. Connecticut, for example, assumes background concentrations of 0 in deriving limits, while Colorado uses actual data. All other things being equal, the discharge limits established by Connecticut will be less stringent than those set by Colorado whenever the actual background concentration is greater than 0.

F3 EPA'S OVERSIGHT OF PERMITTING POLICIES IS LIMITED, BUT NEW INITIATIVES MAY PROMOTE GREATER CONSISTENCY

The EPA, through its regional offices, periodically reviews the states' water quality standards; if it determines that the standards are inconsistent with the requirements of the Clean Water Act—because, for example, the standards do not adequately protect the designated uses of the water or are not scientifically defensible—it disapproves them. However, the EPA does not consistently review policies that could significantly affect the implementation of the standards in permits, either when a state submits its standards for approval or when an EPA regional office reviews individual permits before they are issued.

As a result of an apparent inconsistency in the EPA's regulations, some states are not including the relevant implementation policies when they submit their water quality standards to the EPA for review and approval.[15] According to the regulations, the states must submit to the EPA for review information on the designated uses of their waters and the numeric or narrative criteria for specific pollutants as well as 'information on general policies' that may affect the application or implementation of the standards. However, the EPA's regulations also provide that the states may exercise discretion over what general policies they include in their standards. In EPA's regions I and VI, for example, program officials believe that: (1) the states are under no obligation to submit their implementation policies, such as their policies on considering background concentration, for the EPA's review, and (2) the EPA cannot require the states to do so. Officials at EPA's headquarters and regional offices acknowledge that there is some confusion about what information the states must submit for review.

EPA officials maintain that even if the agency has not reviewed the states' implementation policies in the course of approving the standards, it can control the use of these policies when the EPA's regional offices review individual permits and have the opportunity to disapprove those permits that do not adequately protect water quality. However, on average the EPA's regional offices review only about 10% of the permits issued to major facilities by the 40 states authorized to issue permits. Moreover, the EPA is considering a new initiative that will eliminate reviews of permits before issuance and will instead provide

---

[15]According to the Chief of the EPA's Water Quality Standards Branch, if the EPA has an opportunity to review a state's implementation policies and determines that a policy would prevent the state's water quality standards from being achieved, the agency will disapprove the state's standards.

for postissuance reviews of a sample of permits. According to the Acting Director of the EPA's Permits Division, such reviews are a better use of the EPA's resources because they require less staff time and EPA's reviewers will not be pressured to meet deadlines for public comment. However, he said that, as a general rule, the EPA will not reopen permits. Thus, identified problems may not be addressed until the permits come up for renewal, usually every 5 years. If the EPA becomes aware of a significant problem, the regional office will work with the applicable state to attempt to remedy the situation.

Because the EPA relies on its regional offices to oversee the states' implementation policies, it does not maintain national information on these policies. Moreover, except for some efforts by its regional offices, the EPA has not assessed the impact of the differences among the states. EPA headquarters officials told us that, although such an assessment might be useful, they have no plans to conduct one, in part because they do not have the resources or a specific legislative requirement to do so. In some instances, the EPA's regional offices have tried to identify and resolve differences in the states' implementation policies because they have been concerned about the extent of these differences.[16] However, some states have resisted these initiatives on the basis that they should not be required to comply with policies that are not required nationwide.

The EPA is considering regulatory changes that could enhance the agency's ability to monitor the states' implementation policies. According to a March 1995 draft of an advance notice of proposed rule making, the EPA plans to solicit comments on, among other things, the kind of information on implementation policies that the states should be required to submit for the EPA's approval. In the case of mixing zones, for example, the EPA is seeking comments on whether the states should be required to describe their methods for determining the location, size, shape, and other characteristics of the mixing zones that they will allow. The Chief of the EPA's Water Quality Standards Branch told us that, although other priorities could postpone the rule making, the EPA has not revised the applicable regulations since 1983, and some changes are therefore needed.

While potential regulatory changes are as yet undefined, the Office of Water has embarked on a strategy for watershed management that could, by itself, achieve greater consistency among the states' NPDES programs, including the standards and policies the states use to derive the discharge limits for the facilities within the same watershed. Watershed management means identifying all sources of pollution and integrating controls on pollution within hydrologically defined drainage basins, known as watersheds. Under this approach, all of the stakeholders in a watershed's area—including federal, state, and local regulatory

---

[16]For example, in 1994 EPA's Region VIII drafted a regional policy on mixing zones because it was concerned about inconsistencies in the approaches used by the states in the region and perceived inadequacies in the national guidance.

authorities; municipal and industrial dischargers; other potential sources of pollution; and interested citizens—agree on how best to restore and maintain water quality within the watershed.

In March 1994, the Permits Division of the EPA's Office of Water published its NPDES Watershed Strategy to describe the division's plans for incorporating the NPDES program's functions into the broader watershed management approach. Although the strategy does not specifically discuss interstate watersheds, EPA officials believe that the states will identify such areas and, where reasonable, coordinate the issuance of NPDES permits. EPA officials believe that as a practical matter, the watershed management approach will cause the states to resolve differences in their standards and implementation policies as they attempt to issue NPDES permits consistently in shared water bodies and watersheds.

F4 THE ENVIRONMENTAL PROTECTION AGENCY'S AND THE STATES' RESPONSIBILITIES

Figure F2 illustrates the roles and responsibilities of the EPA and the state agencies in developing water quality standards and implementing them in the permits issued to municipal and industrial wastewater treatment facilities under the NPDES program.

The EPA issues guidance on water quality criteria[17] for specific pollutants that the states may use in developing numeric criteria for their water quality standards. States may also use other data to develop their numeric criteria as long as these criteria are scientifically defensible. The states' water quality standards—and any policies that affect the implementation of these standards—are subject to the EPA's approval.

In determining whether water-quality-based controls are warranted, the states' and the EPA's permitting authorities: (1) analyze a facility's wastewater to identify the type and amount of pollutants being discharged, and (2) determine whether these levels of pollutants will cause, have a 'reasonable potential' to cause, or will contribute to causing the facility's discharge to exceed the state's water quality criteria. This assessment has one of three possible effects on a

---

[17]The EPA's water quality criteria consist of technical information on the effects of pollutants or chemicals on water quality, including the water's physical, chemical, biological, and aesthetic characteristics. Such criteria address the effects of pollutants not only on surface waters but also on sediment, the wildlife that feeds on aquatic life within the waters, and other aspects of the water ecosystem. As we reported in 1994, the EPA has issued criteria to protect human health for 91 of the 126 priority pollutants and criteria to protect aquatic life for 30 of these pollutants. For additional information on the status of the EPA's efforts to develop water quality criteria, see our report *Water Pollution: EPA Needs to Set Priorities for Water Quality Criteria Issues* GAO/RCED-94-117, June 17, 1994.

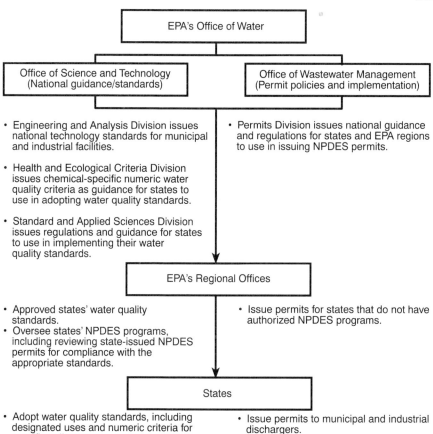

Figure F2 Roles and responsibilities of EPA and States in developing NPDES discharge limits.

facility's permit; it may result in: (1) a discharge limit, if the amount of pollutants being discharged violates, is likely to violate, or will contribute to violating the criteria that protect the receiving waters; (2) a requirement for monitoring to gather additional data in order to determine whether a limit is warranted; or (3) neither a limit nor a monitoring requirement, if the amount of pollutants being discharged will not violate, is unlikely to violate, or will not contribute to violating the criteria that protect the receiving waters.

Appendix G

Superfund: How States Establish and Apply Environmental Standards When Cleaning up Sites

UNITED STATES GENERAL ACCOUNTING OFFICE
GAO/RCED-96-70FS March 20, 1996

Pursuant to a congressional request, the General Accounting Office (GAO) provided information on how states establish and apply environmental standards when cleaning up Superfund sites, focusing on whether states: (1) base their standards on human health risks, and (2) provide flexibility so that the level of cleanup can be adjusted according to the extent of contamination.

GAO found that: (1) 20 of the 21 states reviewed base their hazardous waste site standards on the danger posed to human health, and the cost and technical feasibility of achieving them; (2) states base their groundwater standards on existing federal drinking water standards; (3) when states set their environmental standards at levels other than the federal limit, they tend to be more stringent; (4) states provide more flexibility in adjusting the cleanup level when the cleanup involves soil pollution rather than groundwater pollution, in order to reflect a particular site's condition and health risk; (5) more than half of the states with soil standards regularly allow their cleanup levels to be adjusted for site-specific conditions; (6) less than one-fourth of the states with groundwater standards allow their cleanup levels to be adjusted; and (7) those states not allowing cleanup level adjustments view their groundwater as a potential source of drinking water and implement different standards, depending on the projected use of land or groundwater.

G1 BACKGROUND

Under the Comprehensive Environmental Response, Compensation, and Liability Act (CERCLA), which created the Superfund program in 1980, the Environmental Protection Agency (EPA) assesses uncontrolled hazardous waste sites and places those posing the greatest risks to human health and the environment on the National Priorities List (NPL) for cleanup. As of September 1995, this list included 1232 sites.

Cleanup standards and the degree of cleanup needed for Superfund sites are discussed in section 121(d) of the CERCLA statute, as amended by the

This appendix is a US Government work and, as such, is in the public domain in the United States of America.

Superfund Amendments and Reauthorization Act of 1986 (SARA). This section requires that Superfund sites be cleaned up to the extent necessary to protect both human health and the environment. In addition, cleanups must comply with requirements under federal environmental laws that are legally 'applicable or relevant and appropriate' (ARAR) as well as with such state environmental requirements that are more stringent than the federal standards. Furthermore, Superfund cleanups must at least attain levels established under the Safe Drinking Water Act and the Clean Water Act, where such standards are relevant and appropriate as determined by the potential use of the water and other considerations.

The federal standards most frequently considered relevant and appropriate for groundwater cleanups at Superfund sites are set under the Safe Drinking Water Act. This act establishes standards, called maximum contaminant levels (MCL), for certain contaminants in water delivered by public drinking water systems. As of March 1996, the MCLs included numeric limits on about 70 contaminants. The MCLs take into account estimates of the human health risks posed by contaminants. They also consider whether it is technically and economically feasible to reduce the contamination to a level that no longer poses a health risk. Although MCLs are legally applicable to drinking water systems, section 121(d) of CERCLA generally requires that they be considered relevant and appropriate standards for cleaning up contaminated groundwater that is a potential source of drinking water. For example, the MCL for benzene is 5 μg/L. This concentration would generally be the cleanup level for benzene in groundwater that is a potential source of drinking water unless the state has promulgated a more stringent standard or other requirement that is relevant and appropriate.

There are few federal standards for contaminants in soil that are considered potentially applicable or relevant and appropriate except those for certain highly toxic contaminants, most notably polychlorinated biphenyls (PCB) and lead. Under the Toxic Substances Control Act (TSCA), EPA sets requirements for cleaning up PCB contamination. In addition, the EPA has issued guidance for cleaning up lead in soil.

Early in its investigation of a site, the EPA determines, on the basis of the contamination present and the conditions at the site, which chemical-specific and other standards may be considered applicable or relevant and appropriate. As the EPA proceeds with the selection of a cleanup method, it adjusts the list of standards to be considered on the basis of information gained during its investigation. Among the potential standards considered are any state environmental standards that are more stringent than the federal standards for the same contaminants.

In addition to numeric standards for specific contaminants, some states have set more generalized standards or policies that may have to be considered when cleaning up Superfund sites. For example, some states have established 'antidegradation' policies for groundwater that could require more stringent cleanups than cleanups based on health risks. These policies are intended, among other things, to protect the state's groundwater as a potential source of drinking water.

If federal or state standards do not exist for a given contaminant, the party responsible for cleaning up a Superfund site may use a site-specific risk assessment to help establish a cleanup level for that contaminant. A risk assessment evaluates the extent to which people may be exposed to the contaminant, given its concentration and the physical characteristics of the site. For example, the type of soil and the depth of the groundwater may affect whether and how quickly waste will migrate and reach a population. A risk assessment uses exposure and toxicity data to estimate the increased probability, or risk, that people could develop cancer or other health problems through exposure to this contamination. A risk estimate can be used along with the proposed waste management strategy to help determine the extent of the cleanup necessary at a site.

The EPA has published guidance for conducting risk assessments, a set of documents referred to collectively as the Risk Assessment Guidance for Superfund. These documents outline the well-established risk assessment principles and procedures that can be used to gather and assess information on human health risks. The documents also include information on mathematical models that can be used to estimate health risks at a site, given the contaminants present and the means of exposure to them. In addition to this guidance, EPA maintains an Integrated Risk Information System (IRIS), an on-line database on the toxicity of numerous chemicals, and publishes the *Health Effects Assessment Summary Tables* (HEAST), another source of information on contaminants' toxicity. The EPA uses this guidance in conducting baseline risk assessments at Superfund sites, which it uses in deciding whether the human health and environmental risks posed by the contaminants are serious enough to warrant cleaning up the sites. Some states also use the EPA's risk assessment guidance in setting their standards for specific chemicals.

G2 THE BASES STATES USED IN DEVELOPING STANDARDS

States that have set environmental standards have made decisions about what levels, or concentrations, of chemical contaminants can remain at hazardous waste sites after cleanups. We analyzed the processes that the states in our survey said they went through, as well as the factors that they said they took into consideration, in developing their soil and groundwater standards. In this section, we summarize: (1) the extent to which the states based their soil standards on estimates of the human health risks posed by contaminants at the sites, and (2) the methods that the states used to estimate these risks. We then report on the factors other than health risks that the states said they considered when developing their soil standards. As the bases for the states'

Table G1   Number of states in our survey with their own cleanup standards

Type of standards	Number of states[a]
States with either soil or groundwater standards or with both types of standards	21
States with soil standards	13
States with groundwater standards	20

[a]Our survey included 33 states, which encompass 91% of the sites on the EPA's list of the most contaminated sites in the nation.

standards for groundwater differed somewhat from those for soil, we summarized the information on groundwater standards separately. Finally, as federal drinking water standards are frequently used as cleanup standards for groundwater, we compared the states' groundwater standards with the federal standards for the same contaminants to determine the extent of their correspondence.

We have included the information we obtained from the 33 states in our survey. In all, 21 of the 33 states had set their own standards for either soil or groundwater, or for both media.[1] (See Table G1.)

Thirteen of the 21 states had set their own soil standards, and 20 had set some groundwater standards that were in addition to or different from the MCLs for drinking water, as discussed in the remainder of this section.

G2.1 CONSIDERATION OF HEALTH RISKS IN SETTING SOIL STANDARD

All 13 of the states with soil standards indicated that they considered risks to human health when developing their standards. The number of chemical-specific standards per state ranged from about 10 to nearly 600. All but one of these states generally relied on the EPA's guidance for estimating health risks from contaminants (Missouri had developed its soil standards before the EPA issued its guidance). These states said that they had used the EPA's guidance, either alone or in combination with their own methodologies and policies, to estimate health risks (see Table G2). For example, Pennsylvania said that it had used the EPA's guidance to estimate the toxicity of contaminants and its own model to estimate how much contamination from the soil might travel into

---

[1]This report represents the status of the states' standards at the time of our survey. For example, several states explained that they had modified their standards within the last year, and six told us that they were developing or considering whether to develop soil standards. EPA officials working with the states on cleanups pointed out that some states' standards have evolved and are less stringent now than they were formerly.

Table G2 Health considerations used by 13 states in developing their own soil standards

Health considerations	Number of states
Use of the EPA's guidance to estimate human health risk	
States used the EPA's guidance only	3
States used both the EPA's guidance and their own methodology	9
State developed its standards before the EPA issued its guidance	1
Level of carcinogenic risk that states would allow at a site after cleanup[a]	
States used a 1 in 100 000 increased probability that exposure would cause cancer	5
States used a 1 in 1 000 000 increased probability that exposure would cause cancer	8
Concentration of contaminants causing other health effects that states would allow to remain on site	
States used a measure equivalent to the EPA's	11
States used a more stringent measure than the EPA	2

[a]EPA uses a range of risk levels between 1 in 10 000 and 1 in 1 million; the states all used risk levels within this range.

groundwater. These estimates are two of the major components in the health risk calculation.

Even after a cleanup, contaminants remaining at a site will pose some level of risk to human health. Therefore, the states, when setting standards for contaminants in soil, had to decide what level of risk they would allow to remain after a cleanup. For contaminants that can cause cancer, or carcinogens, the states had to determine an acceptable level of risk, or increased probability, that an individual would develop cancer from a lifetime's exposure to the remaining contamination. All 13 states generally chose risk levels that fell within the range of increased probability that the EPA uses at Superfund sites, which extends from 1 in 10 000 to 1 in 1 000 000. As shown in Table G2, eight states chose the more stringent risk level of 1 in 1 000 000 for individual carcinogens in soil, while five states chose the somewhat less stringent risk level of 1 in 100 000.[2] For noncarcinogens in soil, 11 states used the same measure that the EPA uses at Superfund sites, while two states used a somewhat more stringent measure.

---

[2]Some states chose different risk levels for different contaminants. For example, a state may have used a 1 in 1 000 000 risk level for a contaminant linked by strong evidence to cancer in humans, while using a 1 in 100 000 risk level for other carcinogens. In these cases, we categorized the state as generally using a 1 in 1 000 000 risk level.

G2.2 CONSIDERATION OF OTHER FACTORS IN SETTING SOIL STANDARDS

Ten of the 13 states considered factors in addition to health risk when setting their soil standards. As a result, their standards could be either more or less stringent than those based solely on estimates of health risks. These other factors included the following:

1. *Chemical levels that occur naturally in the environment.* In some locations, certain contaminants may exist naturally in the soil in concentrations differing from those that would be allowed under standards based on risks to human health. For such contaminants, the states typically set their standards at the naturally occurring levels rather than at the levels based solely on risk. In some cases, this practice would result in less stringent cleanups than would be necessary to meet the risk-based standards. However, as some chemicals do not occur naturally in the environment, this practice would in some instances result in more stringent cleanups than would otherwise be required.
2. *Detection limits and practical quantification limits.* When the concentrations of some contaminants that could remain in the soil without posing health risks fell below the levels that can be accurately measured or detected by current technology, the states said that they typically adopt less stringent, but measurable, concentrations as their standards.
3. *Secondary, or aesthetic, criteria.* Some chemicals cause unpleasant odors or other problems at levels that do not pose human health risks. The states may set their standards for these chemicals below risk-based levels to protect the public from such problems.

G2.3 CONSIDERATION OF HEALTH RISKS IN SETTING GROUNDWATER STANDARDS

Twenty of the 33 states we surveyed said that they had set some chemical-specific standards that would limit the concentrations of various toxic

Table G3 Health risk as the basis for states' groundwater standards in states surveyed

States' use of health risk estimates to develop standards	Number of states
States had set some of their own groundwater standards	20
States developed their own risk estimates for some of their standards	16
States predominantly adopted others' standards based on health risks	3
State had no historical information on the development of its standards	1

Table G4 Health considerations used by 13 states in developing their own groundwater standards

Health considerations	Number of states
Use of the EPA's guidance to estimate human health risk	
States used the EPA's guidance only	7
States used both the EPA's guidance and their own methodology	9
Level of carcinogenic risk that states would allow at a site after cleanup[a]	
States used a 1 in 100 000 increased probability that exposure would cause cancer	3
States used a 1 in 1 000 000 increased probability that exposure would cause cancer	13
Concentrations of contaminants causing other health effects that states would allow to remain on site	
States used a measure equivalent to the EPA's	15
States used a more stringent measure than the EPA's	1

[a]EPA uses a range of risk levels between 1 in 10 000 and 1 in 1 million; the states all used risk levels within this range.

chemicals which could be present in groundwater at Superfund sites. These states not only adopted some of the existing federal standards, such as MCLs, but also set some standards in addition to or different from them. The number of chemical-specific standards per state ranged from about 30 to nearly 600. While the remaining states that we surveyed had not developed any of their own groundwater standards, the federal MCLs are typically used as Superfund cleanup standards for groundwater.

Nineteen of the 20 states had based their groundwater standards, at least in part, on estimates of the human health risks posed by exposure to chemical contaminants (see Table G3).

In the remaining state, none of the officials currently involved in implementing the standards could provide historical information on how the standards had been developed. Sixteen of the states had calculated their own health risk estimates when setting the standards for at least some of the contaminants. Three of the states had not predominantly developed their own estimates but had instead adopted standards developed by others, including some or all of the MCLs, that were based on estimates of health risks.

All 16 states that had developed formulas for calculating human health risks had used guidance from the EPA on how to estimate such risks, either alone or in combination with their own procedures and formulas (see Table G4). In setting their standards, 13 of these states used a risk level of 1 in 1 000 000 for individual carcinogens, while three states used the less stringent risk level of 1 in 100 000. For individual noncarcinogens, 15 states used a measure that was as stringent as the EPA's, while one state used a more stringent measure.

G2.4 CONSIDERATION OF OTHER FACTORS IN SETTING GROUNDWATER STANDARDS

All but two of these 16 states said that they had considered factors in addition to human health risks when setting their groundwater standards. Taking such factors into account can affect the concentration of a chemical that a state will allow to remain under its standard. As a result, a standard may be either more or less stringent than one based solely on human health risks.

For groundwater as for soil, these factors included the levels of chemical contaminants that occur naturally in the environment and secondary, or aesthetic, factors, such as the taste and color of drinking water. They also included the following:

1. *Cost and technical feasibility.* If achieving the cleanup level required by a standard based on human health risk would be too costly or technically infeasible, then some states would set the standard at a less stringent level. Similarly, the EPA takes cost and technical feasibility into account when setting some federal standards.
2. *Groundwater protection policies/laws.* Some states have adopted conservative laws or policies to protect groundwater as a potential source of drinking water. For example, some states have 'antidegradation' policies that, for chemicals that do not occur naturally in the environment, may require more stringent cleanups than would be required solely on the basis of risk.

G2.5 DIFFERENCES BETWEEN THE STATES' GROUNDWATER STANDARDS AND THE FEDERAL MAXIMUM CONTAMINANT LEVELS

Because the federal MCLs are typically used as cleanup standards for groundwater used as drinking water at Superfund sites and many of the states based some of their own groundwater standards on the federal MCLs, we compared the states' standards for contaminants with the corresponding MCLs. We found that if a federal MCL existed for a chemical that was included in a state's standards, the state usually set its standard at this level. However, a majority of the states had standards for a few chemicals that differed from the MCLs. These standards tended to be more stringent than the MCLs.

The states offered a variety of explanations for why their standards were more stringent than the federal MCLs. Two states set more stringent levels for certain contaminants if they could detect the contaminants at levels below the MCLs. Several states reported that some of their standards were more stringent because these standards had not been adjusted, as the MCLs had been, for other factors, such as cost or technical feasibility.

Some states' standards may also have been more stringent because the states had antidegradation policies for groundwater. For example, Wisconsin mandates that the environment be restored to the extent practicable. Consequently, it has set 'preventive action limits' for contaminants in groundwater that may be used to determine the extent of the cleanup required at Superfund sites unless it can be shown that meeting such limits would not be technically or economically feasible. All of the preventive action limits are more stringent than the corresponding federal MCLs. They limit the concentrations of chemicals that can cause cancer to one-tenth the concentrations allowed under the MCLs, and they limit the concentrations of chemicals that can cause other health effects to one-fifth the concentrations allowed under the MCLs. However, the state allows exemptions for contaminants that occur naturally at levels exceeding the preventive action limits.

Nearly all of the states had only a few, if any, standards for contaminants that were less stringent than the corresponding federal MCLs. However, under SARA, only those numeric standards that are more stringent than the federal standards are to be considered as cleanup levels at Superfund sites.

G3 THE FLEXIBILITY STATES PROVIDED IN APPLYING STANDARDS TO CLEANUPS

Even though the states have set environmental standards, they have found that applying these standards uniformly to all sites may not be effective because conditions can vary from one hazardous waste site to another. As a result, sites may pose different levels of health risks and may, therefore, require different degrees of cleanup. We examined whether the states: (1) allow the level of cleanup determined to be necessary under their standards to be adjusted to take into account site-specific conditions, and (2) set different standards for different uses of the land or groundwater (e.g., set more stringent cleanup standards for land that could be used for residential than for industrial purposes). Overall, the states provided more flexibility in applying their soil standards than their groundwater standards.

G3.1 FLEXIBILITY TO ADJUST CLEANUP LEVELS BASED ON SOIL STANDARDS TO ACCOUNT FOR SITE-SPECIFIC CONDITIONS

Eight of the 13 states that had soil standards indicated that they allow the extent of the cleanup deemed necessary under their standards to be adjusted for site-specific factors. For example:

1. Georgia's risk reduction standards include the option of determining cleanup target concentrations for contaminants on the basis of site-specific risk assessments.
2. Minnesota characterized its standards as 'quick reference numbers' rather than fixed limits, that are considered when determining how extensively

to clean up a site. Thus, cleanup levels can be tailored to local conditions. For example, if exposure to contaminants in soil were reduced or eliminated because the soil was inaccessible, the cleanup levels would not need to meet the standards. Alternatively, if multiple contaminants with the same toxic effect were found at the same location, the cleanup level for each individual contaminant might be more stringent than the standard.

3. Pennsylvania said that it has developed interim standards pending final regulations for about 100 soil contaminants, but considers these to be 'worst case' numbers that can be adjusted to reflect site-specific conditions.

In contrast, the remaining five states said that, in general, their soil standards were fixed limits on the concentrations of contaminants that must be met when cleaning up a site. While adopting relatively inflexible standards, these five states said that they do provide for certain exceptions under limited circumstances. For example, one state mentioned that it did not require that soil lying beneath a building be cleaned up to comply with the standards because the building would prevent people's exposure to the contaminated soil. Alternatively, under certain conditions, some states allow cleanups to be based on site-specific risk assessments. Three of these states also said that they permitted less stringent cleanup levels than those based on their standards if meeting them was not technologically feasible or if naturally occurring levels of chemicals in the local environment were higher than the levels set by the standards. However, the use of such alternatives was the exception rather than the rule.

Some of the states also indicated that even if they do not provide much flexibility in applying their standards, they may permit flexibility in determining how to achieve the required level of protection. For example, instead of requiring costly incineration of contaminated soil to meet its standards, a state may allow the area to be covered with a clay cap so that people cannot come into contact with the contaminants.

G3.2 DIFFERENT SOIL STANDARDS FOR DIFFERENT LAND USES OR EXPOSURES

The states may also provide flexibility by establishing different standards for different projected uses of the land at a site. Ten of the 13 states with soil standards told us they had set such standards. For example, Michigan said that it had defined soil standards for three types of land uses: residential, industrial, and commercial (with two subcategories). Generally, the more stringent standards apply to residential property, as people are more likely to be exposed to contaminants for a longer period of time on residential property than on other types of property.

G3.3 FLEXIBILITY TO ADJUST CLEANUP LEVELS BASED ON GROUNDWATER STANDARDS TO ACCOUNT FOR SITE-SPECIFIC CONDITIONS

While most states allowed flexibility in their cleanup levels for soil, the states were less flexible in setting cleanup levels for groundwater. The degree of flexibility largely depended on whether the groundwater was considered a potential source of drinking water.

Only four of the 20 states with groundwater standards said that they regularly allowed the cleanup levels for groundwater used as drinking water to be based on site-specific considerations. Texas, for example, incorporated this option in one of its cleanup strategies.

Texas reported that it has three alternative 'risk reduction' standards, any one of which can be used to clean up a site, with the state's approval. Cleanups under the first of these standards must ensure that the levels of contamination are no higher than the levels that occur naturally in the environment. Cleanups under the second standard must meet fixed limits set by the state, including federal MCLs that the state has adopted, and place a notice in deed records to inform future property owners of any contamination left on the property. Cleanups under the third standard must also use federal MCLs when available, but for contaminants without corresponding MCLs, site-specific risk-based cleanup levels can be determined on the basis of the site's projected use. The third standard also requires deed notification.

The remaining 16 states indicated that, in general, for groundwater used as drinking water or considered potentially usable as drinking water, their standards were fixed limits that must be achieved during cleanup. Most of these states did say, though, that they allowed certain limited exceptions to their standards or the use of a site-specific risk assessment under some circumstances. For example, if the contaminated water came from an area where the contamination would not immediately threaten communities, a state might let the contamination be reduced naturally over time rather than require that it be cleaned up immediately.

The states gave various reasons for the relative inflexibility of their groundwater standards for drinking water. First, some of the states said that they were mirroring the federal MCLs for drinking water, which are also fixed limits. Some of the states also noted that they consider groundwater that may possibly be used as drinking water as a valuable resource that needs to be conserved.

G3.4 DIFFERENT STANDARDS FOR DIFFERENT USES OF GROUNDWATER

Although the states in our survey told us that their standards for groundwater used as drinking water are relatively fixed, some states also reported that they provided some degree of flexibility by not classifying all groundwater as drinking water. They also set less stringent standards for groundwater that would not be considered a potential source of drinking water. For example, Connecticut's groundwater classification system acknowledges that in certain areas, such as those that have had long-term industrial or commercial use, the groundwater would not be a suitable source of drinking water unless it was treated. The state does not usually require that the groundwater in such areas be cleaned up to the standards for drinking water if it has a high mineral content or if it is located in a geological formation that does not yield much water.

Twelve of the 20 states with groundwater standards said that they set different standards for different current or potential uses of groundwater. Thus, these states set separate standards for groundwater used for agricultural purposes, groundwater of special ecological significance (e.g., supporting a vital wetland), and groundwater in urban, industrial, or commercial areas.[3]

Seven of these 12 states indicated that site-specific factors can be taken into account when determining the extent of the cleanup needed for these other types of groundwater. For example, Rhode Island told us that it allows the cleanup levels for some contaminants to differ from the levels set in its standards. For example, vapors escaping from volatile organic chemicals in the groundwater could accumulate in overlying buildings and cause potential health effects. In some cases, these vapors could build up and cause threats of explosion. In setting its 'urban' groundwater standards, this state conservatively assumed that the building would not be ventilated and that the vapors from the underlying groundwater would be trapped in the buildings. However, in deciding how extensively to clean up a site, the state allows for a consideration of site-specific factors, such as depths to groundwater. When site-specific factors are considered, the cleanup levels may not need to be as stringent as the standards alone would require.

---

[3]While another five states exempted some of their groundwater from meeting drinking water standards, these states did not set other standards for the exempt water.

Index